CAMBRIDGE ENGINEERING SERIES

GENERAL EDITOR:

SIR JOHN BAKER, F.R.S.

THE ELASTIC ANALYSIS
OF FLAT GRILLAGES

T0275593

THE ELASTIC ANALYSIS OF FLAT GRILLAGES

WITH PARTICULAR REFERENCE TO SHIP STRUCTURES

BY

J. CLARKSON
B.SC., A.M.R.I.N.A.

Principal Scientific Officer at the
Naval Construction Research Establishment
Dunfermline

CAMBRIDGE
AT THE UNIVERSITY PRESS
1965

CAMBRIDGE UNIVERSITY PRESS
Cambridge, New York, Melbourne, Madrid, Cape Town, Singapore, São Paulo, Delhi

Cambridge University Press
The Edinburgh Building, Cambridge CB2 8RU, UK

Published in the United States of America by Cambridge University Press, New York

www.cambridge.org
Information on this title: www.cambridge.org/9780521046732

First published 1965
This digitally printed version 2008

A catalogue record for this publication is available from the British Library

Library of Congress Catalogue Card Number: 65–16200

ISBN 978-0-521-04673-2 hardback
ISBN 978-0-521-09067-4 paperback

PREFACE

This book describes important developments in the elastic analysis of flat grillages, as applied to ship structures, during the past twenty years. Historically, problems in ship grillages were solved many years ago, and the contributions by Professor H. A. Schade of the University of California, and Dr G. Vedeler of Det Norske Veritas, Oslo, are of particular importance. More recently, the subject has received much attention, particularly regarding the methods of numerical analysis, the effect of plating, and the formulation of design data sheets. In naval applications, much of the work has been carried out at the Naval Construction Research Establishment, Dunfermline, and this forms the main subject of the present text, reference also being made to work in the civil and structural engineering fields where appropriate.

I would like to thank a number of colleagues who have taken part in the work, particularly Mr S. Kendrick who was responsible for the computer program using the displacement method described in chapter 5, and Mr D. F. Holman who developed the finite Fourier series method in chapter 4, from earlier work by Dr J. M. Klitchieff of Belgrade University. Much encouragement has also been received from Dr J. L. King, Chief Scientist at the Naval Construction Research Establishment. The text is being published by permission of the Navy Department of the Ministry of Defence.

J. C.

October 1964

CONTENTS

NOTATION

In general, the notation is given as it is introduced in the text, but the following standard symbols are not defined:

E = Young's modulus

G = Shear modulus

μ = Poisson's ratio

A = area of cross-section of beam

I = moment of inertia of beam

Z = section modulus of beam = (moment of inertia)/(distance from neutral axis to outermost fibre)

R = intersection reaction

w = downwards deflection

x and y denote orthogonal directions in the plane of the grillages

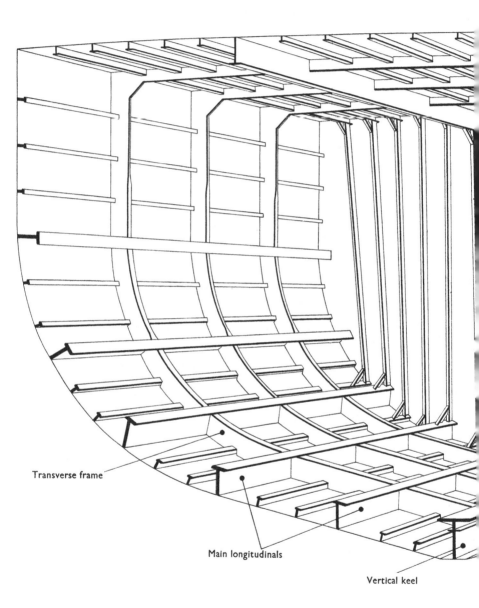

Transverse frame

Main longitudinals

Vertical keel

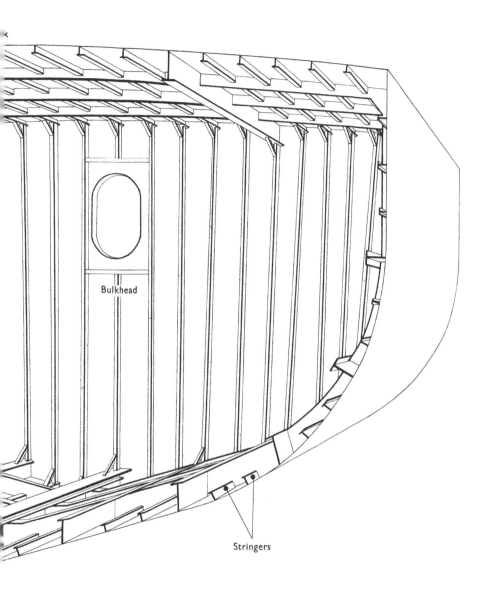

Bulkhead

Stringers

Fig. 1. Portion of main hull structure of frigate.

Chapter 1

INTRODUCTION

It is now 18 years since Vedeler published his treatise entitled *Grillage Beams in Ships and Similar Structures*, and the period has seen some important developments in the elastic analysis of this type of structure. This is particularly the case for the grillages in ship structures, and the object of the present text is to describe the advances which are most important to the designer. Carefully conducted experiments have been carried out on plated grillages, leading to simple rules for determining the contribution of the plating to the strength of the beams. This period has also seen the introduction of the digital electronic computer, and the present trend is to adopt straightforward numerical methods which were previously rejected due to the prohibitively lengthy arithmetic involved. Nowadays, with very high-speed computers readily available, these methods are often the most popular of all.

Grillage is the term given to a structure of intersecting beams which is loaded normal to its surface. Grillages are particularly common in ship structures, and a typical hull construction for a frigate is shown in fig. 1. The beams are the stiffening members for the plating which is present to provide watertight integrity, and they are usually placed longitudinally and transversely forming a mesh which intersects orthogonally, at right angles. The decks, bottoms and bulkheads of ships are all examples of flat grillages. This form of construction is also commonly used in the decks of bridges. This text is therefore mainly concerned with the analysis of flat plated grillages of orthogonally intersecting beams; it should be mentioned that the term 'beam' is applied to either the longitudinal or transverse stiffening members, not particularly to the transverse beams of ships' decks, a term used in naval architecture. Some of the methods described are also applicable to non-orthogonal grillages.

A grillage is a highly redundant structure which cannot be analysed purely by statical considerations, and, to obtain a solution, recourse has to be made to the conditions for compatible deflections of the components of the structure. For many redundant structures, plastic analysis methods have been advocated as providing a simpler yet more realistic approach than elastic analysis. In this method, the load for final plastic collapse is calculated, and the working load is taken as some fraction of this. To calculate the collapse load for a grillage, the positions of the plastic 'hinges', where the beam section has become completely plastic, are postulated, and sufficient hinges are assumed to allow collapse as a mechanism. A simple application of the principle of virtual work leads to the collapse load, without any reference to the sequence of events leading up to collapse, and the mechanism giving to the smallest collapse load is the correct one. This collapse load is independent of any initial residual stresses due to fabrication.

Unfortunately, the methods of plastic analysis, which have been developed for structures involving bending action such as the portal frames of buildings, are not so obviously applicable to grillages in which the members are rigidly connected at the intersections. This is because a plane structure which is supported at all its edges cannot be laterally deformed into a surface curved in two directions without introducing a membrane stretching action. This is the case, even if the supporting structure at the edges cannot provide any membrane restraint, and the action inside the grillage is then one of tension around the centre and tangential compression at the edges. The plating of ship grillages adds appreciably to the membrane strength of these structures. Long after the structure has exhausted its bending strength, the structure will deform as a membrane, and the final failure will usually be at some weak point in the details of the connections at the intersections, but gross plastic deformation may occur long before this. For these reasons, the methods of plastic design are thought to be of more limited application in the case of grillages, and the present text is concerned with elastic methods of design. Elastic design ensures that the range of stress due to the load is kept to some fraction of the yield stress, and in this way all noticeable plastic deformation is prevented. The only yielding possible is on the first application of load, due to residual fabrication stresses.

The plating of present-day grillages is usually welded to the stiffening members, and, to calculate the maximum stresses and deflections of these members, the approach adopted is to consider the plating as an effective flange to both longitudinal and transverse beams. In this way, the analysis is reduced to that of an unplated grillage, and is merely a question of applying Euler–Bernoulli beam theory to a structure containing a large number of redundancies. The early chapters of this text are concerned with the basic equations of this analysis, and their solution by approximate methods designed to

[2]

reduce the arithmetic, by a neat exact method using Fourier series, and by straightforward numerical methods using a modern digital computer. These latter methods are the most suitable for including the effects of shear deflection and torsional rigidity, which may be regarded as refinements. The problem of assigning an effective breadth for the beams in plated ship grillages is then discussed, and methods for designing simple welded connections are described, the relevant experimental data from various research projects being presented. The last chapters are concerned with design data sheets which have been worked out for certain cases of concentrated and uniform pressure loadings, and the question of minimum weight design. The design of grillages, as distinct from analysis, has so far received comparatively little attention from research workers, but we may now regard ourselves as standing at a turning point: with digital computers commonly available, the problems of analysis are completely solved, and these computers can be readily utilized to study some of the important problems of minimum weight design.

Chapter 2

FORCE METHOD

The elastic analysis of a plane grillage of beams under loads normal to the surface consists of satisfying the conditions of equilibrium and compatibility at every intersection point. Thus at a typical intersection point there is a reaction on the longitudinal and an equal and opposite reaction on the transverse member. This intersection reaction is statically indeterminate, but may be derived from the condition that, at the intersection, both beams are attached and have equal deflections. In a grillage containing p longitudinal and q transverse members, there are $p \times q$ intersections, and pq unknown reactions. The deflections may be calculated in terms of these reactions, and by equating the longitudinal and transverse beam deflection at every intersection, pq linear simultaneous equations are obtained, which may be solved numerically. The bending moments are then obtained by statics. It will be assumed for the present that shear deflections and torsional rigidity of the beams may be neglected.

2.1. Numerical example 3 × 1 grillage

We will consider first the action of the grillage in fig. 2 under equal loads of magnitude P at positions P_1 and P_2. Let the upwards reaction on the longitudinal at positions 1 and 2 be R_1 and R_2 respectively. There will be equal and opposite reactions on the transverse members at these positions. The intersection deflections may now be derived in terms of forces using simple beam theory, as follows:

$$\text{Transverse beam deflection at position 1} = \frac{a^3}{EI_a} \frac{R_1}{48}. \tag{2.1}$$

$$\text{Transverse beam deflection at position 2} = \frac{a^3}{EI_a} \frac{R_2}{48}. \tag{2.2}$$

Longitudinal beam deflection at position 1

$$= \frac{b^3}{EI_b} \left(-\frac{R_1}{48} - \frac{11R_2}{768} + \frac{41P}{24 \times 64} \right). \tag{2.3}$$

Longitudinal beam deflection at position 2

$$= \frac{b^3}{EI_b} \left(-\frac{11R_1}{384} - \frac{R_2}{48} + \frac{39P}{16 \times 64} \right). \tag{2.4}$$

[4]

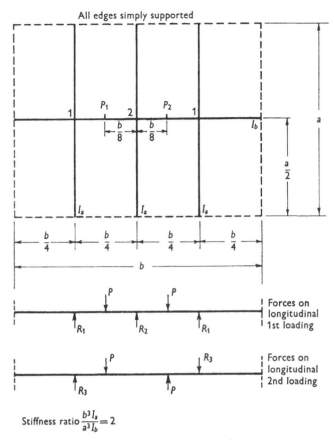

Fig. 2. 3 × 1 grillage example.

Equating (2.1) and (2.3), (2.2) and (2.4), we obtain two simultaneous equations in R_1 and R_2. Introducing the numerical value for the stiffness ratio $b^3 I_a / a^3 I_b = 2$, and solving,

$$R_1 = 0 \cdot 510P \quad \text{and} \quad R_2 = 0 \cdot 752P.$$

Consider, now, the action of the grillage under a downward load P at position P_1 and an upward load P at P_2. The intersection reaction at position 2 will be zero by symmetry. Let the reaction at 1 be R_3 acting upward on the left end of the longitudinal and downward on the right end.

Transverse beam deflection at position 1 (left) $= \dfrac{a^3}{EI_a} \dfrac{R_2}{48}$.

[5]

Longitudinal beam deflection at position 1 (left)

$$= \frac{b^3}{EI_b}\left(-\frac{R_3}{384} + \frac{11P}{96 \times 64}\right).$$

Equating these two expressions and introducing the numerical stiffness ratio, we obtain

$$R_3 = 0{\cdot}137P.$$

2.2. Use of symmetry of structure

Applying the principle of superposition to the above example and adding the solutions for the two loadings, we may derive the solution for a load $2P$ at P_1 and zero at P_2. This unsymmetrical loading might have been discussed directly in terms of three unknown reactions at the intersection points, leading to three simultaneous equations. By making use of the symmetry of the structure, and splitting the load into symmetrical and anti-symmetrical components, we have reduced

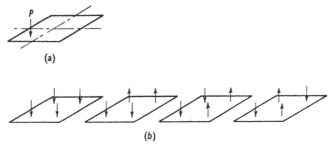

(a)

(b)

Fig. 3. Treatment of unsymmetrical loadings. (a) Two axes of symmetry, unsymmetrical loading. (b) Four symmetrical components of loading for grillage calculation.

the problem to independent sets of two and one simultaneous equations. This may seem trivial for the very small grillage considered, but it is of major importance to reduce the labour of numerical solution of the equations, in larger grillages. In general, any unsymmetrical loading on a grillage having two axes of symmetry may be resolved into four components having differing types of symmetry, according to the scheme illustrated in fig. 3. Then, only the intersection points in one quarter of the grillage appear in the analysis, the remainder being given by symmetry. In future discussions of symmetrical grillages, we will usually refer only to the number of beams in one quarter of the grillage, including the central beams where present.

2.3. Flexibility coefficients

The process of setting up equations in the forces may be largely reduced to a numerical operation, using general formulae for the flexibility of a beam under concentrated loadings. These coefficients β_{st} are defined by the relation

$$w_s = \beta_{st} \frac{a^3}{EI} R_t,$$

where w_s is the deflection at a distance x_s from the end supports under a load R_t at distance x_t from the end, a = span of beam and EI = flexural rigidity. The coefficients are derived from beam theory, and for uniform beams are given in the following table of expressions, valid for $x_s \leqslant x_t$. The coefficients for $x_s > x_t$ may be derived using Maxwell's reciprocal theorem.

TABLE 1. Flexibility coefficients

(a) Unsymmetrical load R_t, distance x_t from one end

   (i) simply supported ends $\beta_{st} = \dfrac{1}{6}\dfrac{x_s}{a}\left(1-\dfrac{x_t}{a}\right)\left(\dfrac{2x_t}{a}-\dfrac{x_s^2+x_t^2}{a^2}\right)$

   (ii) clamped ends $\qquad\beta_{st} = \dfrac{1}{6}\left(\dfrac{x_s}{a}\right)^2\left(1-\dfrac{x_t}{a}\right)^2\left[3\dfrac{x_t}{a}-\dfrac{x_s}{a}\left(1+\dfrac{2x_t}{a}\right)\right]$

(b) Symmetrical loading: two loads R_t, x_t from each end

   (i) simply supported ends $\beta_{st} = \dfrac{1}{6}\dfrac{x_s}{a}\left[\dfrac{3x_t}{a}\left(1-\dfrac{x_t}{a}\right)-\dfrac{x_s^2}{a^2}\right]$

   (ii) clamped ends $\qquad\beta_{st} = \dfrac{1}{6}\left(\dfrac{x_s}{a}\right)^2\left[3\dfrac{x_t}{a}\left(1-\dfrac{x_t}{a}\right)-\dfrac{x_s}{a}\right]$

(c) Anti-symmetrical loading: downwards load R_t, x_t from left end, and upwards load R_t, x_t from right end

   (i) simply supported ends $\beta_{st} = \dfrac{1}{6}\dfrac{x_s}{a}\left(1-\dfrac{2x_t}{a}\right)\left(\dfrac{x_t}{a}-\dfrac{x_s^2+x_t^2}{a^2}\right)$

   (ii) clamped ends $\qquad\beta_{st} = \dfrac{1}{6}\left(\dfrac{x_s}{a}\right)^2\left(1-\dfrac{2x_t}{a}\right)\left[3\dfrac{x_t}{a}\left(1-\dfrac{x_t}{a}\right)\right.$

$$\left.-\dfrac{x_s}{a}\left(1+2\dfrac{x_t}{a}-2\left(\dfrac{x_t}{a}\right)^2\right)\right]$$

2.4. Range of application

Solution by the Force Method requires first the formulation of a set of linear simultaneous equations, and for uniform beams this is simple and straightforward, using algebraic formulae. For non-uniform beams, this part of the analysis would be appreciably more difficult and laborious. The equations may be solved on a desk mechanical calculating machine for up to about five unknowns. For larger numbers of unknowns, a digital electronic computer would normally be used. Since present medium sized computers have standard subroutines which can solve up to about 70 equations, the Force Method may now be regarded as a very powerful means of grillage analysis.

Chapter 3

APPROXIMATE METHODS

Although the Force Method has become very attractive recently
due to the introduction of digital electronic computers, more approximate methods, designed to avoid the solution of large sets of equations,
may be of interest to those who do not have ready access to a computer.
Historically, these approximate methods were of considerable importance, and three methods will now be described briefly and
applied numerically to the simple 9 × 3 beam grillage shown in fig. 4.
The beams of each set are equal and evenly spaced, all edges are
simply supported, and the grillage is acted on by a central concentrated load. An exact solution, derived by solving the simultaneous
equations, gives the following results:

central deflection = $0 \cdot 000731 Pa^3/EI_g$,
maximum longitudinal girder bending moment = $0 \cdot 0330 Pa$,
maximum transverse stiffner bending moment = $0 \cdot 0242 Pa$,

3.1. Minimum Potential Energy Method

In this method, a deflected shape is assumed

$$w = \sum_{m=1}^{\lambda} \sum_{n=1}^{\lambda} a_{mn} X_m(x) Y_n(y). \tag{3.1}$$

The functions $X_m(x)$, $Y_n(y)$ should satisfy the boundary conditions at
the edges of the grillage, and it simplifies the subsequent algebra if
they also satisfy the conditions,

$$\int_0^a X''_{m_1} X''_{m_2} dx = 0 \quad \text{for} \quad m_1 \neq m_2,$$

$$\int_0^b Y''_{n_1} Y''_{n_2} dy = 0 \quad \text{for} \quad n_1 \neq n_2.$$

The numerical coefficients a_{mn} are determined by the condition that
the change in potential energy due to the assumed deflections is a
minimum. The potential energy may be written,

$$V = V_g + V_s - W,$$

[9]

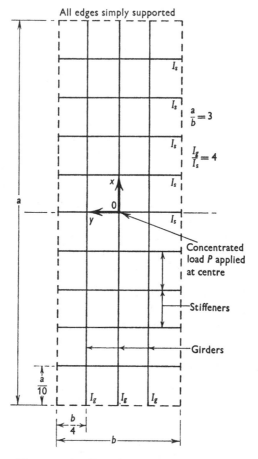

Fig. 4. 9 × 3 grillage for approximate analysis.

where V_g, V_s are the elastic strain energies of the girders and stiffeners and W is the work done by the external load. For minimum potential energy

$$\frac{\partial V}{\partial a_{mn}} = 0 \quad \text{for} \quad m = 1, 2, \dots, \lambda \quad \text{and} \quad n = 1, 2, \dots, \lambda, \qquad (3.2)$$

which provides λ^2 simultaneous linear equations for the λ^2 unknown coefficients a_{mn}. In contrast to the simultaneous equations of the Force Method, the energy equations can usually be iterated.

[10]

The strain energy of the ith girder is given by

$$V_{gi} = \tfrac{1}{2}EI_g \int_0^a \left(\frac{\partial^2 w}{\partial x^2}\right)^2_{y=y_i} dx = \tfrac{1}{2}EI_g \sum_{m=1}^{\lambda} M_m c_{mi}^2,$$

where $\quad M_m = \displaystyle\int_0^a X_m''^2 \, dx \quad$ and $\quad c_{mi} = \displaystyle\sum_{n=1}^{\lambda} a_{mn} Y_m(y_i),$

i.e.
$$V_g = \tfrac{1}{2}EI_g \sum_{i=1}^{g} \sum_{m=1}^{\lambda} M_m c_{mi}^2. \tag{3.3}$$

Similarly for the stiffeners,

$$V_s = \tfrac{1}{2}EI_s \sum_{i=1}^{s} \sum_{n=1}^{\lambda} N_n b_{in}^2, \tag{3.4}$$

where $\quad N_n = \displaystyle\int_0^b Y_n''^2 \, dy \quad$ and $\quad b_{in} = \displaystyle\sum_{m=1}^{g} a_{mn} X_m(x_i).$

Differentiating equations (3.3) and (3.4), we obtain

$$\frac{\partial V_g}{\partial a_{mn}} = EI_g M_m \sum_{r=1}^{\lambda} a_{mr} g_{rn}, \tag{3.5}$$

$$\frac{\partial V_s}{\partial a_{mn}} = EI_s N_n \sum_{r=1}^{\lambda} a_{rn} s_{mr}, \tag{3.6}$$

where
$$g_{mn} = \sum_{i=1}^{g} Y_m(y_i) \, Y_n(y_i),$$

$$s_{mn} = \sum_{i=1}^{s} X_m(x_i) \, X_n(x_i).$$

The work done by the external load is given by

$$W = \int_0^a \int_0^b \left[\sum_{m=1}^{\lambda} \sum_{n=1}^{\lambda} a_{mn} X_m Y_n \right] q(x,y) \, dx \, dy,$$

where $q(x, y)$ is the load per unit area,

i.e.
$$\frac{\partial W}{\partial a_{mn}} = \int_0^a \int_0^b q(x,y) X_m Y_n \, dx \, dy. \tag{3.7}$$

Introducing (3.5), (3.6) and (3.7) into (3.2), we have λ^2 simultaneous equations for the λ^2 unknown coefficients a_{mn}, which may be expressed in matrix form as follows:

$$EI_g[M][A][G] + EI_s[S][A][N] = [K]. \tag{3.8}$$

[11]

The matrices are all square of order $\lambda \times \lambda$ and the elements are defined as

$$m_{mn} = 0 \quad \text{for} \quad m \neq n, \qquad m_{mm} = M_m = \int_0^a X_m''^2\, dx,$$

$$n_{mn} = 0 \quad \text{for} \quad m \neq n, \qquad n_{mm} = N_n = \int_0^b Y_n''^2\, dy,$$

$$g_{mn} = \sum_{i=1}^{g} Y_m(y_i)\, Y_n(y_i), \qquad s_{mn} = \sum_{i=1}^{s} X_m(x_i)\, X_n(x_i),$$

$$k_{mn} = \int_0^a \int_0^b q(x,y)\, X_m(x)\, Y_n(y)\, dx\, dy.$$

In the limit, as very many terms are included and $\lambda \to \infty$, the solution given by (3.8) will approach the exact solution, provided $X_m(x)$, $Y_n(y)$ are sufficiently general to allow the true deflected shape to be expressed as an infinite double series in $X_m(x)$, $Y_n(y)$.

The functions $X_m = \sin(m\pi x/a)$, $Y_n = \sin(n\pi y/b)$ satisfy the boundary conditions for a simply supported grillage. Introducing these into (3.8), we obtain

$$\frac{\pi^4 E I_g}{2a^3}[P]^4[A][G] + \frac{\pi^4 E I_s}{2b^3}[S][A][P]^4 = [K], \tag{3.9}$$

where

$$P_{mn} = 0 \quad \text{for} \quad m \neq n, \quad P_{mm} = m,$$

$$g_{mn} = \sum_{i=1}^{g} \sin\frac{im\pi}{g+1}\sin\frac{in\pi}{g+1},$$

$$s_{mn} = \sum_{i=1}^{s} \sin\frac{im\pi}{s+1}\sin\frac{in\pi}{s+1},$$

$$k_{mn} = \int_0^a \int_0^b q(x,y)\sin\frac{m\pi x}{a}\sin\frac{n\pi y}{b}\, dx\, dy;$$

$$\left.\begin{array}{l} g = \text{numbers of girders} \\ s = \text{number of stiffeners} \end{array}\right\} \text{ assumed to be evenly spaced.}$$

To evaluate g_{mn}, we use the expression

$$g_{mn} = \frac{1}{4}\left[\frac{\sin(2g+1)(m-n)\pi/(2g+2)}{\sin(m-n)\pi/(2g+2)} - \frac{\sin(2g+1)(m+n)\pi/(2g+2)}{\sin(m+n)\pi/(2g+2)}\right],$$

which gives
$$g_{mn} = \tfrac{1}{2}(g+1)$$

for $m - n = 2k_1(g+1)$ and $m + n \neq 2k_2(g+1)$ simultaneously,

$$g_{mn} = -\tfrac{1}{2}(g+1)$$

for $m - n \neq 2k_1(g+1)$ and $m + n = 2k_2(g+1)$ simultaneously,

[12]

provided that $m \neq k_1(g+1)$ and $n \neq k_2(g+1)$, where k_1 and k_2 are two integers.

$g_{mn} = 0$ for all other cases.

s_{mn} is obtained from the same formulae with g replaced by s.

For the particular grillage in fig. 4, the equations (3.9) were solved with $\lambda = 9$. For the symmetrical loading considered, the even co-efficients are identically zero, so that this gave a total of 25 unknown coefficients. The equations are well conditioned and may be solved very quickly by iteration, the answers for each cycle being obtained in order of ascending mn using the latest values for all other coefficients in the equation being used. Starting from zero values, the following answers were obtained for the central deflection:

first iteration: $0.0072808Pa^3/EI_g$,
second iteration: $0.00072756Pa^3/EI_g$.

The steady value, obtained after four iterations, was $0.00072751Pa^3/EI_g$. The values for bending moment, obtained by differentiating the expression for deflection, were as follows:

maximum girder bending moment $= 0.0268Pa$,
maximum stiffener bending moment $= 0.0224Pa$.

These values differ by up to 15 % from the exact answers, which is attributable to small inaccuracies in the deflections which show up to a greater extent in the derivatives. Differentiation introduces multiples of m or n to the a_{mn} coefficients, thus affecting convergence of the series.

For a grillage with clamped edges under a symmetrical loading, the functions $X_m = \sin^2(m\pi x/a)$, $Y_n = \sin^2(n\pi y/b)$ satisfy all the necessary conditions. The general form of the analysis is similar to that for simple supports, though more laborious.

3.2. Distributed Reaction Method

This method is applicable to grillages having a very few beams in one direction (the girders) and a large number of equal and evenly spaced beams in the orthogonal direction (the stiffeners). Any external loading applied to the girders may be considered. The essence of the method is that the discrete reactions applied to the girders (counted positive downwards) are replaced by an equivalent distributed reaction, and this will be illustrated for a grillage containing a single

[13]

girder. Considering first a typical stiffener, the intersection deflections may be obtained from its flexibility, giving

$$w = -\beta \frac{b^3}{EI_s} R, \qquad (3.10)$$

using the coefficients in table 1. The reactions on the girders, R, are considered to be equivalent to a uniform load per unit length, $[R(s+1)]/a$, i.e.

$$EI_g \frac{d^4w}{dx^4} = \frac{R(s+1)}{a}, \qquad (3.11)$$

and eliminating R from (3.10) and (3.11), we obtain

$$\frac{EI_g ab^3 \beta}{(s+1) EI_s} \frac{d^4w}{dx^4} + w = 0. \qquad (3.12)$$

Additional terms appear for the externally applied loading. Equation (3.12) is a linear differential equation in w which may be solved by the usual methods, giving four constants of integration which are determined by the boundary conditions. For a fairly large number of stiffeners, this method would involve less arithmetic than the Force Method.

The Distributed Reaction Method will now be presented for a grillage having a central girder and two symmetrically placed side girders under a central concentrated load, with a view to obtaining numerical results for the grillage in fig. 4. The central girder will be denoted by the suffix 0 and the side girders by suffix 1. Considering, first, the flexibility of the stiffeners, we have

$$\left.\begin{aligned}
w_0 &= \frac{b^3}{EI_s} (-\beta_{00} R_0 - \beta_{01} R_1), \\
w_1 &= \frac{b^3}{EI_s} (-\beta_{10} R_0 - \beta_{11} R_1).
\end{aligned}\right\} \qquad (3.13)$$

For the girders under their respective distributed reactions, we have

$$\left.\begin{aligned}
EI_{g0} \frac{d^4w_0}{dx^4} &= \frac{R_0(s+1)}{a}, \\
EI_{g1} \frac{d^4w_1}{dx^4} &= \frac{R_1(s+1)}{a}.
\end{aligned}\right\} \qquad (3.14)$$

Eliminating the reactions R_0 and R_1, we obtain the following simultaneous linear differential equations:

$$a_0 w_0^{iv} + b_0 w_1^{iv} + w_0 = 0, \qquad (3.15)$$

$$a_1 w_0^{iv} + b_1 w_1^{iv} + w_1 = 0, \qquad (3.16)$$

where
$$a_0 = \frac{EI_{g1}ab^3\beta_{00}}{(s+1)EI_s}, \quad a_1 = \frac{\beta_{10}}{\beta_{00}}a_0,$$

$$b_0 = \frac{EI_{g1}ab^3\beta_{01}}{(s+1)EI_s}, \quad b_1 = \frac{\beta_{11}}{\beta_{01}}b_0.$$

Equations (3.15) and (3.16) may be combined to obtain an eighth-order equation in w_0:

$$(b_1a_0 - a_1b_0)\frac{d^8w_0}{dx^8} + (a_0+b_1)\frac{d^4w_0}{dx^4} + w_0 = 0. \qquad (3.17)$$

The solution may be written

$$w_0 = (C_1e^{\alpha x} + C_2e^{-\alpha x})\sin\alpha x + (C_3e^{\alpha x} + C_4e^{-\alpha x})\cos\alpha x$$
$$+ (D_1e^{\gamma y} + D_2e^{-\gamma x})\sin\gamma x + (D_3e^{\alpha x} + D_4e^{-\gamma x})\cos\gamma x, \qquad (3.18)$$
where

$$\alpha^4 = \tfrac{1}{2}(a_0 + b_1 + r), \quad \gamma^4 = \tfrac{1}{2}(a_0 + b_1 - r), \quad r = (a_0 - b_1)^2 + 4a_1b_0,$$

and $C_1, C_2, ..., D_4$ are constants of integration.

Using equations (3.15), (3.16) and (3.18), an expression may be derived for the deflection of the side girders:

$$w_1 = k_0[(C_1e^{\alpha x} + C_2e^{-\alpha x})\sin\alpha x + (C_3e^{\alpha x} + C_4e^{-\alpha x})\cos\alpha x]$$
$$+ k_1[(D_1e^{\gamma x} + D_2e^{-\gamma x})\sin\gamma x + (D_3e^{\gamma x} + D_4e^{-\gamma x})\cos\gamma x], \qquad (3.19)$$
where $\quad k_0 = (-a_0 + b_1 + r)/2b_0, \quad k_1 = (-a_0 + b_1 - r)/2b_0.$

For simply supported edges and a central concentrated load, the boundary conditions are as follows:

$$x = 0: \quad \frac{dw_0}{dx} = \frac{dw_1}{dx} = 0, \quad EI_g\frac{d^3w_0}{dx^3} = \frac{P}{2}, \quad EI_g\frac{d^3w_1}{dx^3} = 0.$$

$$x = \frac{a}{2}: \quad w_0 = w_1 = \frac{d^2w_0}{dx^2} = \frac{d^2w_1}{dx^2} = 0.$$

Using the expressions (3.18) and (3.19) for w_0 and w_1, we obtain:

$$C_1 = \tfrac{1}{2}F(\sin\mu - \cos\mu - \cosh\mu + \sinh\mu)/(\cos\mu + \cosh\mu),$$

$$C_2 = -F - C_1,$$

$$C_3 = \tfrac{1}{2}F(\sin\mu + \cos\mu + \cosh\mu - \sinh\mu)/(\cos\mu + \cosh\mu),$$

$$C_4 = -F + C_3,$$

where $F = k_1P/8E\alpha^3(k_0 - k_1)I_g, \mu = \alpha a$. The values of the D constants are similar to those for the C's, replacing α, k_0 and k_1 by γ, k_1 and k_0, respectively.

The girder bending moments are obtained by differentiating equations (3.18) and (3.19). The stiffener bending moments are obtained by statics from discrete reactions at the intersections, evaluated from deflections using the relations (3.13).

For the grillage in fig. 4, equations (3.15) and (3.16) become

$$8w_0^{iv} + 11w_1^{iv} + 320 \times 3^4 w_0/a^4 = 0,$$

$$11w_0^{iv} + 16w_1^{iv} + 640 \times 3^4 w_1/a^4 = 0,$$

giving

$$\alpha = 4 \cdot 5018/a,$$

$$\gamma = 13 \cdot 0725/a,$$

$$k_0 = -k_1 = 1/\sqrt{2} = 0 \cdot 70711,$$

$$C_1 = 5 \cdot 9533 \times 10^{-6} Pa^3/EI_g, \quad D_1 = 0 \cdot 000022774 \times 10^{-6} Pa^3/EI_g,$$

$$C_2 = 679 \cdot 11 \times 10^{-6} Pa^3/EI_g, \quad D_2 = 27 \cdot 978 \times 10^{-6} Pa^3/EI_g,$$

$$C_3 = 8 \cdot 9781 \times 10^{-6} Pa^3/EI_g, \quad D_3 = -0 \cdot 000079815 \times 10^{-6} Pa^3/EI_g,$$

$$C_4 = 694 \cdot 04 \times 10^{-6} Pa^3/EI_g, \quad D_4 = 27 \cdot 978 \times 10^{-6} Pa^3/EI_g.$$

The deflections and bending moments at the centre of the grillage are as follows:

$$\text{central deflection} = 0 \cdot 000731 Pa^3/EI_g,$$

$$\text{girder bending moment} = 0 \cdot 0369 Pa,$$

$$\text{stiffener bending moment} = 0 \cdot 0244 Pa.$$

The deflection agrees identically with the exact solution to three significant figures, but the girder bending moment is out by 12 % due to replacing the stiffeners by a continuous medium. The stiffener bending moment is more exact.

3.3. Orthotropic Plate Method

In this method, the beam structure is considered to be represented by a continuous medium having an equivalent stiffness per unit width in the two orthogonal directions. It is only applicable for uniform grillages in which the beams of both sets are equal and evenly spaced, and to obtain reasonably accurate results, there should also be a fair number of beams in each direction. Having represented the structure by an orthotropic plate, the formal solutions of flat plate analysis may be adopted. The method is discussed here for unplated grillages, though in this instance it is a simple extension to allow for the biaxial stresses in the plating of plated grillages.

[16]

Moment carried by a single girder $= EI_g \dfrac{\partial^2 w}{\partial x^2}$.

Spacing of girders $= \dfrac{b}{(g+1)}$.

If the girders are spaced closely enough, they may be replaced by plating which carries a moment per unit width,

$$M_x = EI_g \frac{\partial^2 w}{\partial x^2} \frac{(g+1)}{b}. \tag{3.20}$$

Similarly, the stiffeners may be replaced by plating which carries a moment per unit width

$$M_y = EI_s \frac{\partial^2 w}{\partial y^2} \frac{(s+1)}{a}. \tag{3.21}$$

Letting q be equal to the pressure acting on the surface of the grillage, equilibrium of an element of the equivalent orthotropic plate gives rise to the equation

$$\frac{\partial^2 M_x}{\partial x^2} + \frac{\partial^2 M_y}{\partial y^2} = q,$$

and introducing (3.20) and (3.21) we obtain

$$\frac{(g+1)EI_g}{b} \frac{\partial^4 w}{\partial x^4} + \frac{(s+1)EI_s}{a} \frac{\partial^4 w}{\partial y^4} = q, \tag{3.22}$$

which is a linear differential equation with constant coefficients. It may be generalized and rewritten non-dimensionally as

$$A \frac{\partial^4 w}{\partial X^4} + B \frac{\partial^4 W}{\partial X^2 \partial Y^2} + C \frac{\partial^4 W}{\partial Y^4} = p, \tag{3.23}$$

where, for the particular case of an unplated grillage considered here,

$$W = w/a, \quad X = x/a, \quad Y = y/b, \quad p = a^3 b q/(g+1) EI_g,$$

$$A = 1, \quad B = 0, \quad C = \frac{a^3(s+1)I_s}{b^3(g+1)I_g}.$$

In the Navier solution of the plate equation (3.23), the loading is expressed as a double Fourier series which leads to a very simple solution. Unfortunately, a concentrated load implies an infinite pressure and the series does not converge, so that this solution is not generally applicable. The method of solution to be described here was introduced by Levy, and will be described for a simply supported grillage under a central concentrated load.

The deflection is assumed to be given by the series

$$W = \sum_{m=1}^{\infty} F_m(Y) \sin m\pi X. \tag{3.24}$$

Introducing this into the differential equation gives

$$\sum_{m=1}^{\infty} [m^4 \pi^4 A F_m(Y) - m^2 \pi^2 B F_m''(Y) + C F_m^{\mathrm{iv}}(Y)] \sin m\pi X = 0.$$

This equation is satisfied for all X if

$$C F_m^{\mathrm{iv}} - m^2 \pi^2 B F_m'' + m^4 \pi^4 A F_m = 0$$

for each value of m. Solving this ordinary differential equation in the usual way and substituting into (3.24) we obtain

$$W = \sum_{m=1}^{\infty} \left[W_m \sin \frac{m\pi d\,Y}{2} \sinh \frac{m\pi c\,Y}{2} + X_m \sin \frac{m\pi d\,Y}{2} \cosh \frac{m\pi c\,Y}{2} \right.$$
$$\left. + Y_m \cos \frac{m\pi d\,Y}{2} \sinh \frac{m\pi c\,Y}{2} + Z_m \cos \frac{m\pi d\,Y}{2} \cosh \frac{m\pi c\,Y}{2} \right] \sin m\pi X, \tag{3.25}$$

where W_m, X_m, Y_m and Z_m are constants of integration,

$$c^2 = \sqrt{(a^2+b^2)} + a, \quad d^2 = \sqrt{(a^2+b^2)} - a, \quad a = B/C,$$
$$b = (1/C)\sqrt{(4AC - B^2)}.$$

To the complementary solution (3.25) we have to add particular integrals to describe the loading. For a concentrated load at the centre, the grillage is divided into three regions as illustrated in fig. 5. The

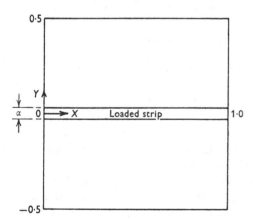

Fig. 5. Axes for non-dimensional analysis by orthotropic plate method.

[18]

region $0.5 > Y > 0.5\alpha$ has no external load and the complementary solution (3.25) gives the complete solution. The region

$$0.5\alpha > Y > -0.5\alpha$$

is called the 'loaded strip', though its width α is infinitely small for a concentrated load. A particular integral may be derived by considering this strip as a simply supported beam under the externally applied loading, If we define the non-dimensional point load $P = a^2 Q/(g+1)\,EI_g$, where $Q =$ dimensional point load, the particular integral may be written

$$W = \frac{2P}{\alpha A\pi^4} \sum_{m=1,\,3,\,\ldots}^{\infty} \frac{(-1)^{\frac{1}{2}(m-1)}}{m^4} \sin m\pi X, \qquad (3.26)$$

and the complete solution, using the symmetry in Y, is

$$W = \frac{2P}{\alpha A\pi^4} \sum_{m=1,\,3,\,\ldots}^{\infty} \left[W_{m_1} \sin\frac{m\pi d\,Y}{2} \sinh\frac{m\pi c\,Y}{2} \right.$$
$$\left. + Z_{m_1} \cos\frac{m\pi d\,Y}{2} \cosh\frac{m\pi c\,Y}{2} + \frac{(-1)^{\frac{1}{2}(m-1)}}{m^4} \right] \sin m\pi X. \quad (3.27)$$

It is convenient to write the solution for $0.5 > Y > 0.5\alpha$ in the form

$$W = \frac{2P}{\alpha A\pi^4} \sum_{m=1,\,3,\,\ldots}^{\infty} \left[W_{m_1} \sin\frac{m\pi d\,Y}{2} \sinh\frac{m\pi c\,Y}{2} \right.$$
$$\left. + X_{m_1} \sin\frac{m\pi d\,Y}{2} \cosh\frac{m\pi c\,Y}{2} + Y_{m_1} \cos\frac{m\pi d\,Y}{2} \sinh\frac{m\pi c\,Y}{2} \right.$$
$$\left. + Z_{m_1} \cos\frac{m\pi d\,Y}{2} \cosh\frac{m\pi c\,Y}{2} \right] \sin m\pi X. \quad (3.28)$$

The solution for $-0.5\alpha > Y > -0.5$ is identical to (3.28) by symmetry. Only the terms in odd m appear in (3.27) and (3.28) due to the form of the particular integral (for loadings symmetrical about $X = 0.5$).

The constants of integration are determined by the following boundary conditions:

$$W = \frac{\partial^2 W}{\partial Y^2} = 0 \quad \text{for} \quad Y = 0.5,$$

$$W, \quad \frac{\partial W}{\partial Y}, \quad \frac{\partial^2 W}{\partial Y^2}, \quad \frac{\partial^3 W}{\partial Y^3} \quad \text{continuous at } Y = 0.5\,\alpha.$$

This gives rise to the following expressions:

$$W_{m_2} = W_{m_1} + \frac{(-1)^{\frac{1}{2}(m-1)}}{m^4}\,\frac{(c^2-d^2)}{2cd},$$

$$Z_{m_2} = Z_{m_1} - \frac{(-1)^{\frac{1}{2}(m-1)}}{m^4},$$

$$W_{m_1} = -\frac{\pi\alpha(c^2+d^2)\,(-1)^{\frac{1}{2}(m-1)}}{8cdm^3}$$

$$\times \left\{\frac{c\sinh{(m\pi c/4)}\cosh{(m\pi c/4)} + d\sin{(m\pi d/4)}\cos{(m\pi d/4)}}{\cosh^2{(m\pi c/4)} - \sin^2{(m\pi d/4)}}\right\},$$

$$X_{m_1} = \frac{\pi\alpha(c^2+d^2)\,(-1)^{\frac{1}{2}(m-1)}}{8dm^3},$$

$$Y_{m_1} = \frac{\pi\alpha(c^2+d^2)\,(-1)^{\frac{1}{2}(m-1)}}{8cm^3},$$

$$Z_{m_1} = \frac{\pi\alpha(c^2+d^2)\,(-1)^{\frac{1}{2}(m-1)}}{8cdm^3}$$

$$\times \left\{\frac{d\sinh{(m\pi c/4)}\cosh{(m\pi c/4)} - c\sin{(m\pi d/4)}\cos{(m\pi d/4)}}{\cosh^2{(m\pi c/4)} - \sin^2{(m\pi d/4)}}\right\},$$

neglecting terms of second-order magnitude in α. It will be noticed that, on substituting these constants into (3.27) and (3.28), the term α, the width of the loaded strip, cancels out.

The expressions for deflection converge as $1/m^3$, and a small number of terms is adequate to obtain numerical values. Differentiating twice to obtain the bending moments results in a series in $1/m$. The convergence of the moments is improved by integrating the curvature over the beam spacing, rather than by adopting the curvature at the actual beam position:

$$M_g = \frac{(g+1)\,EI_g}{a}\int_{Y-\frac{1}{2}\beta}^{Y+\frac{1}{2}\beta}\frac{\partial^2 W}{\partial X^2}\,dY, \qquad (3.29)$$

$$M_s = \frac{(s+1)\,EI_s}{b}\int_{X-\frac{1}{2}\gamma}^{X+\frac{1}{2}\gamma}\frac{\partial^2 W}{\partial Y^2}\,dX, \qquad (3.30)$$

where $\beta = 1/(g+1)$ and $\gamma = 1/(s+1)$.

Substituting for W from (3.28) which may be considered as valid throughout the structure since (3.27) is only required for the infinitely narrow loaded strip, the integral in (3.30) is as follows:

$$\int_{X-\frac{1}{2}\gamma}^{X+\frac{1}{2}\gamma} \frac{\partial^2 W}{\partial Y^2} dX = \frac{P(c^2+d^2)}{16\pi^2 A c d} \sum_{m=1,\,3,\,\ldots}^{\infty} \frac{(-1)^{\frac{1}{2}(m-1)}}{m^2} \Big\{ [(c^2-d^2)\,W_{m_2}$$

$$-2cd\,Z_{m_2}] \sin \frac{m\pi d\,Y}{2} \sinh \frac{m\pi c\,Y}{2}$$

$$+c(c^2+d^2) \sin \frac{m\pi d\,Y}{2} \cosh \frac{m\pi c\,Y}{2}$$

$$+d(c^2+d^2) \cos \frac{m\pi d\,Y}{2} \sinh \frac{m\pi c\,Y}{2}$$

$$+[2cd W_{m_2} + (c^2-d^2)\,Z_{m_2}] \cos \frac{m\pi d\,Y}{2} \cosh \frac{m\pi c\,Y}{2}\Big\}$$

$$\times \{\cos m\pi(X-\tfrac{1}{2}\gamma) - \cos m\pi(X+\tfrac{1}{2}\gamma)\}. \quad (3.31)$$

To evaluate this summation, it should be noted that the quantities in the square brackets quickly converge, as follows:

$$[(c^2-d^2)\,W_{m_2} - 2cd Z_{m_2}] \to -c(c^2+d^2),$$

$$[2cd W_{m_2} + (c^2-d^2)\,Z_{m_2}] \to -d(c^2+d^2),$$

so that the later terms of the series may be written

$$(c^2+d^2)\frac{(-1)^{\frac{1}{2}(m-1)}}{m^2}\Big\{ c \sin \frac{m\pi d\,Y}{2}\Big(\cosh \frac{m\pi c\,Y}{2} - \sinh \frac{m\pi c\,Y}{2}\Big)$$

$$- d \cos \frac{m\pi d\,Y}{2}\Big(\cosh \frac{m\pi c\,Y}{2} - \sinh \frac{m\pi c\,Y}{2}\Big)\Big\}$$

$$\times \{\cos m\pi(X-\tfrac{1}{2}\gamma) - \cos m\pi(X+\tfrac{1}{2}\gamma)\}$$

$$= \frac{(c^2+d^2)}{2}\frac{e^{-\frac{1}{2}m\pi c Y}}{m^2}[d(\sin m\epsilon_1 + \sin m\epsilon_2 - \sin m\epsilon_3 - \sin m\epsilon_4)$$

$$+ c(\cos m\epsilon_1 + \cos m\epsilon_2 - \cos m\epsilon_3 - \cos m\epsilon_4)], \quad (3.32)$$

where

$$\epsilon_1 = \tfrac{1}{2}\pi(d\,Y + 2X - \gamma - 1), \quad \epsilon_2 = \tfrac{1}{2}\pi(d\,Y - 2X + \gamma - 1),$$

$$\epsilon_3 = \tfrac{1}{2}\pi(d\,Y + 2X + \gamma - 1), \quad \epsilon_4 = \tfrac{1}{2}\pi(d\,Y - 2X - \gamma - 1).$$

For positions on the centre line $Y = 0$, the expression (3.32) reduces to

$$d(c^2+d^2)\frac{1}{m^2}(\sin m\epsilon_1 - \sin m\epsilon_3), \quad (3.33)$$

[21]

and the terms may be evaluated by summing an equivalent series containing Bernoulli's numbers B_n,

$$\sum_{m=1,3,\ldots}^{\infty} \frac{\sin m\epsilon}{m^2} = -\frac{\epsilon}{2}\left(\log_e \frac{\epsilon}{2}\right) + \frac{\epsilon}{2} - \frac{\epsilon^3}{8\times 9} - \frac{14\epsilon^5}{32\times 900} - \frac{62\epsilon^7}{128\times 19{,}845} -$$

$$\ldots - \frac{(2^{2n}-2)\,B_n\,\epsilon^{2n+1}}{4n(2n+1)!} + \ldots,$$

which converges much more rapidly than the original series.

The bending moments of the orthogonal set of beams may be evaluated from the expressions for M_s by interchanging the two sets in the numerical work.

For the particular grillage in fig. 4, taking the X axis parallel to the girders, equation (3.23) becomes

$$\frac{\partial^4 W}{\partial X^4} + 16\cdot 88 \frac{\partial^4 W}{\partial Y^4} = p,$$

giving $c = d = 0\cdot 6978$.

For calculating the bending moments of the girders, the X axis is taken parallel to the stiffeners, giving

$$\frac{\partial^4 W}{\partial X^4} + 0\cdot 05926 \frac{\partial^4 W}{\partial Y^4} = p$$

and $c = d = 2\cdot 8663$.

The deflections and bending moments at the centre of the grillage, calculated from equations (3.28) and (3.30), are as follows:

$$\text{central deflection} = 0\cdot 000738 Pa^3/EI_g,$$
$$\text{girder bending moment} = 0\cdot 0360 Pa,$$
$$\text{stiffener bending moment} = 0\cdot 0280 Pa.$$

The deflection only differs marginally from the exact solution, but the bending moments are out by up to 15%, which is due to replacing the structure of a grillage having as few as 9×3 beams by an orthotropic plate.

The condition of clamped edges at $Y = \pm 0\cdot 5$ may be analysed by introducing the appropriate boundary conditions for determining W_{m_1} to Z_{m_2}. For the case of all four edges clamped, it is necessary to consider additional loadings in the form of bending moments in terms of Fourier series applied to each edge, the coefficients being determined from the condition that the total slope arising from these moments and from the normal loading on the grillage must be zero.

[22]

This involves the numerical solution of a set of simultaneous equations equal in number to the terms included in the edge moments. The procedure is described for isotropic plates in *Theory of Plates and Shells* by Timoshenko and Woinowsky-Krieger, page 197. The advantages of having explicit expressions in the Plate Analogy Method are largely lost when all the edges are clamped, and the method is no longer to be recommended.

Chapter 4

FOURIER SERIES METHOD

Although approximate methods may prove useful, they are always somewhat unsatisfactory, since the degree of approximation is usually in doubt. A very useful formal solution, giving an exact result yet with appreciably less numerical labour than the Force Method, is the Fourier Series Method by Holman. This is a development of work by Klitchieff, an infinite series solution being adapted for grillages in which one set of beams are evenly spaced, to give finite series for deflections and bending moments at the intersections of the beams. The numer of terms in the series is not greater than the number of beams in the equally spaced set, and the chief numerical labour consists of solving p independent sets of q simultaneous equations, rather than the pq equations using the Force Method. The method will be introduced in its simplest form, valid when one set of beams is simply supported, the orthogonal set being evenly spaced, and the notation is shown in fig. 6.

4.1. Solution for B beams with supported ends

The deflection of the sth beam of the simply supported set B can be represented by

$$w_s = \sum_{k=1}^{\infty} C_{ks} \sin \frac{k\pi y}{b}, \qquad (4.1)$$

and the bending moment is then

$$M_s = EI_s \frac{d^2 w_s}{dy^2} = -EI_s \sum_{k=1}^{\infty} \left(\frac{k\pi}{b}\right)^2 C_{ks} \sin \frac{k\pi y}{b}. \qquad (4.2)$$

We may also obtain an expression for this moment from the intersection reactions R_{rs} taken positive downwards:

$$M_s = -\sum_{r=1}^{p} y R_{rs} \left(1 - \frac{y_r}{b}\right) + \sum_{r=1}^{p} R_{rs}(y - y_r) U(y - y_r) \qquad (4.3)$$

for the case where there is no external load applied to the B beams, where $\quad U(t) = 0 \quad (t < 0) \quad$ and $\quad U(t) = 1 \quad (t > 0)$.

Equating these two expressions, multiplying by sin $(m\pi y/b)$ and integrating over $(0, b)$ with respect to y,

$$EI_s C_{ms} \left(\frac{m\pi}{b}\right)^2 \left(\frac{b}{2}\right) = \sum_{r=1}^{p} R_{rs} \left(\frac{b}{m\pi}\right)^2 \sin \frac{m\pi y_r}{b}, \qquad (4.4)$$

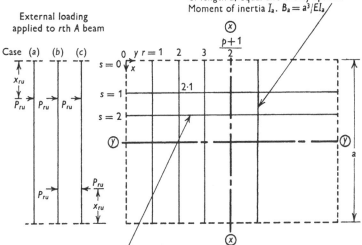

Set A = uniform beams, p in number of length a, equal and evenly spaced. Moment of inertia I_a. $B_a = a^3/EI_a$.

External loading applied to rth A beam

Set B = uniform beams, q in number of length b, not necessarily identical or equally spaced. Moment of inertia of the sth beam = I_s

External loading. Applied to sth B beam

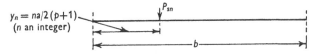

$y_n = na/2(p+1)$
(n an integer)

Fig. 6. Range of grillages and notation.

Loading cases and symmetry: (a) structure of B beams not symmetrical in either position or stiffness about the centre-line yy; no symmetry of loading. (b) Structure of B beams and loading symmetrical about yy; the deflections will be symmetrical about yy, i.e. $w_{r,\,q-s} = w_{rs}$. (c) Structure of B beams symmetric and loading antisymmetric about yy (loading either side of yy of equal magnitude but opposite sign); the deflections will be antisymmetric about yy, i.e. $w_{r,\,q-s} = -w_{rs}$.

Notation:

w_{rs} = downward deflection at beam intersection point rs.

R_{rs} = downward reaction on B beam at intersection point rs.

$\alpha_s = b^3 I_a/48(p+1)^3\, a^3 I_s$.

$$J_m = \cos\frac{m\pi}{2(p+1)}\,\mathrm{cosec}^2\frac{m\pi}{2(p+1)}\left[6\,\mathrm{cosec}^2\frac{m\pi}{2(p+1)}-1\right].$$

$$K_m = \mathrm{cosec}^2\frac{m\pi}{2(p+1)}\left[1+3\cot^2\frac{m\pi}{2(p+1)}\right].$$

C_{ms} = Fourier coefficient for sth B beam.

V_{ms} = parameter formed from C_{ms}; $Z_{ms} = \pi^4 m^4 C_{ms}$.

$$U_t = \sum_{m=1}^{p} Z_{mt}.$$

where m is any positive integer. If the beams of set A are evenly spaced then $hy_r = b_r$ where $h = p+1$ so that equation (4.4) becomes, writing Z_{ms} for $(m\pi)^4 C_{ms}$,

$$EI_s b^{-3} Z_{ms} = 2 \sum_{r=1}^{p} R_{rs} \sin\frac{m\pi r}{h} \quad (m = 1, 2, \ldots). \tag{4.5}$$

Now, for integral values of n and k which are less than h, we have

$$2 \sum_{r=1}^{h-1} \sin\frac{n\pi r}{h} \sin\frac{k\pi r}{h} = h \quad \text{if} \quad n = k, \\ = 0 \quad \text{if} \quad n \neq k. \tag{4.6}$$

Then, from (4.5) we obtain

$$EI_s b^{-3} \sum_{m=1}^{p} Z_{ms} \sin\frac{m\pi r}{h} = 2 \sum_{m=1}^{p} \sum_{t=1}^{p} R_{ts} \sin\frac{m\pi t}{h} \sin\frac{m\pi r}{h}$$
$$= hR_{rs}. \tag{4.7}$$

Thus, the reactions R_{rs} at the joints are given in terms of a finite series of trigonometrical functions Z_{ms} and the bending moments may be found by taking moments.

Expressions may also be derived from the direct calculation of deflections or bending moments at the intersections along the B beams, using the following procedure. The right-hand side of equation (4.5) is a periodic function of m which is unaltered if m be replaced by $v = |m+2jh|$ (j is any integer positive or negative) except for a change of sign if $m+2jh < 0$. Hence

$$Z_{vs} = Z_{ms} \operatorname{sgn}(m+2jh), \tag{4.8}$$

where
$$\operatorname{sgn}(t) = +1 \quad (t > 0), \\ = -1 \quad (t < 0). \tag{4.9}$$

For $y = rb/h$, equation (4.1) becomes

$$w_{rs} = \sum_{k=1}^{\infty} C_{ks} \sin\frac{k\pi r}{h}.$$

Rearranging the terms in this series,

$$w_{rs} = \sum_{m=1}^{p} \sum_{j=-\infty}^{\infty} C_{vs} \sin\frac{m\pi r}{h} \operatorname{sgn}(m+2jh)$$
$$= \sum_{m=1}^{p} \sum_{j=-\infty}^{\infty} m^4 C_{ms}(m+2jh)^{-4} \sin\frac{m\pi r}{h},$$

using (4.8). Now by the known result

$$\sum_{j=-\infty}^{\infty} (m+2jh)^{-4} = \frac{\pi^4}{48 h^4}\left(1 + 3\cot^2\frac{m\pi}{2h}\right)\operatorname{cosec}^2\frac{m\pi}{2h} = \frac{\pi^4}{48h^4} K_m \quad \text{(say),} \tag{4.10}$$

we obtain
$$w_{rs} = \frac{1}{48h^4} \sum_{m=1}^{p} K_m Z_{ms} \sin \frac{m\pi r}{h}. \qquad (4.11)$$

Similarly, for the bending moment in the B beam at joint (r, s),

$$M_{rs} = -EI_s \sum_{k=1}^{\infty} (k\pi/b)^2 C_{ks} \sin(k\pi r/h)$$

$$= -\frac{EI_s}{b^2} \sum_{m=1}^{p} \sum_{j=-\infty}^{\infty} v^2 \pi^2 C_{vs} \sin(m\pi r/h) \operatorname{sgn}(m+2jh)$$

$$= -\frac{EI_s}{b^2} \sum_{m=1}^{p} \sum_{j=-\infty}^{\infty} m^4 \pi^2 C_{ms} (m+2jh)^{-2} \sin(m\pi r/h)$$

$$= -\frac{EI_s}{4b^2 h^2} \sum_{m=1}^{p} Z_{ms} \operatorname{cosec}^2(m\pi/2h) \sin(m\pi r/h) \qquad (4.12)$$

using the summation

$$\sum_{j=-\infty}^{\infty} (m+2jh)^{-2} = \frac{\pi^2}{4h^2} \operatorname{cosec}^2(m\pi/2h). \qquad (4.13)$$

It has been shown that the solution may be expressed in terms of a finite series in Z_{ms}. These coefficients are determined using the condition for compatibility of the deflections. The intersection deflections given by (4.11) may also be derived by considering the forces acting on the A beams:
$$w_{rs} = \frac{a^3}{EI_a} \left(\sum_u \beta_{su} P_{ru} - \sum_{t=1}^{q} \beta_{st} R_{rt} \right), \qquad (4.14)$$

where P_{ru} is a downward external load on the rth A beam at a distance x_{ru} from the end $s = 0$, and the β's are the flexibility coefficients given in table 1, p. 7. Used directly and substituting for w_{rs} and R_{rt} in terms of the Z_{ms} coefficients, this equation is a simultaneous set containing all Z's for $m = 1, 2, ..., p$ and $s = 1, 2, ..., q$. This solution would have no advantage over the Force Method which leads to the same number of simultaneous equations. However, by a process of summation, the equations may be resolved into smaller groups of equations. From (4.14),

$$\sum_{r=1}^{p} w_{rs} \sin \frac{m\pi r}{h} = \frac{a^3}{EI_a} \sum_r \left[\sum_u \beta_{su} P_{ru} - \sum_{t=1}^{q} \beta_{st} R_{rt} \right] \sin \frac{m\pi r}{h}.$$

Substituting for w_{rs} from (4.11) and R_{rt} from (4.7) and evaluating the summations using (4.6),

$$\alpha_s K_m Z_{ms} + \sum_{t=1}^{q} \beta_{st} Z_{mt} = \frac{2b^3}{EI_s} \sum_{r=1}^{p} \sin \frac{m\pi r}{h} \sum_u \beta_{su} P_{ru},$$

where
$$\alpha_s = \frac{b^3 I_a}{48h^3 a^3 I_s}. \qquad (4.15)$$

[27]

The number of simultaneous equations is now equal to q, the number of B beams, and a set is written down and solved for $m = 1, 2, ..., p$ (number of A beams).

4.2. Procedure for numerical solution

Although the detailed analyses by the Fourier Series Method require fairly advanced mathematics, numerical solutions may be obtained for a number of standard problems, without following the detailed mathematical derivations. The general procedure will now be described for three such problems. In all cases, it is necessary for one set of beams, the A beams, to be equal and evenly spaced.

4.2.1. B beams with simply supported ends under any loading on the A beams.

1st Step: calculation of the flexibility coefficients for the A beams, β_{su} and β_{st} in equation (4.14), *using the formulae in table* 1, p. 7. The formulae allowing for symmetry should be used wherever appropriate, and the ranges of the summations in (4.14) are given in table 2. The effect of symmetry is to reduce the number of terms to half. An equation derived from (4.14) should be written down for each intersection point of a typical A beam and for each A beam which has a differing external load.

2nd Step: formulation and solution of equations in Z coefficients. Numerical equations in the coefficients Z_{ms} may now be written down using (4.15), the ranges of the summations in t and u being given in table 2. An equation is written down for each integer value of s over the range given for the summation in t and the set solved for $Z_{m_1}, Z_{m_2}, \ldots$. The process is repeated for each integer value of m from $m = 1$ to $m = p$, inclusive. For loadings symmetrical about xx, $C_{ms} = 0$ for m even.

3rd Step: calculation of deflections, reactions and bending moments. The values at the intersection points are given by equations (4.11), (4.7) and (4.12). Knowing the reactions, the bending moments of the A beams may be found by statics.

4.2.2. B beams with simply-supported ends under concentrated loadings midway between intersections

This case is mainly of interest for analysis under a single concentrated load, which usually causes higher stresses when applied between intersection points. However, the procedure is presented for any num-

TABLE 2. *Ranges of summations*

Symmetry case ...	(i)	(ii)	(iii)
Structure symmetry about centre line yy	No symmetry	Symmetrical	
Loading symmetry about centre line yy	No symmetry	Symmetrical $P_{ru} = P_{r,-u}$	Anti-symmetrical $P_{ru} = -P_{r,-u}$
Ranges of summation: $\sum$ loadings externally u applied to the rth A beam over span	$0 < x < a$	$0 < x \leqslant \tfrac{1}{2}a$	$0 < x < \tfrac{1}{2}a$
$\sum$ integer values of t t over range			
(a) for q odd	1 to q	1 to $\tfrac{1}{2}(q+1)$	1 to $\tfrac{1}{2}(q-1)$
(b) for q even	1 to q	1 to $\tfrac{1}{2}q$	1 to $\tfrac{1}{2}q$

These summation ranges apply to equations (4.15), (4.16), (4.17), (4.21), (4.22) and (4.23).

ber of concentrated loads, provided these are applied along the B beams, mid-way between intersection points, and the simplification due to symmetry is indicated, reducing the numerical work almost by half. In some cases, the effect of a single load may be obtained to sufficient accuracy by considering the combined effect of two symmetrically applied loads, provided these are sufficiently far apart.

1st Step: this is identical with that for loading along the A beams except that there are no terms in the externally applied load, P_{ru}, giving

$$w_{rs} = \frac{a^3}{EI_a} \sum_t \beta_{st} R_{rt}, \tag{4.16}$$

the number of terms in the summation being obtained from table 2. The subsequent analysis differs from that for loading along the A beams.

2nd Step: formulation and solution of equations in V coefficients (parameters formed from C coefficients). The general equation is as follows:

$$\alpha_s \left[K_m V_{ms} + J_m \sum_n P_{sn} \sin \frac{m\pi y_n}{b} \right] = -\sum_t \beta_{st} V_{mt}, \tag{4.17}$$

where P_{sn} is an external load applied at a distance y_n from the end $r = 0$ of the sth B beam. The summation in n covers all the external loads applied to the sth B beam.

$$J_m = \cos \frac{m\pi}{2h} \operatorname{cosec}^2 \frac{m\pi}{2h} \left[6 \operatorname{cosec}^2 \frac{m\pi}{2h} - 1 \right], \tag{4.18}$$

[29]

and the V coefficients are convenient parameters having the dimension of load formed from the Fourier coefficients, C_{ms}. An equation should be written down for each integer value of s over the ranges given for summations in t in table 2 and the set of simultaneous equations solved for V_{m_1}, V_{m_2}, etc. Numerical values of V_{ms} should be evaluated for each integer value of m from $m = 1$ to $m = p$, inclusive. For loadings symmetrical about xx, $V_{ms} = 0$ for m even.

3rd Step: calculation of deflections, reactions and bending moments. At the intersection points

$$w_{rs} = \frac{b^3}{48h^4EI_s} \sum_{m=1}^{p} \left[K_m V_{ms} + J_m \sum_n P_{sn} \sin\frac{m\pi y_n}{b} \right] \sin\frac{m\pi r}{h}, \quad (4.19)$$

$$R_{rs} = \frac{1}{h} \sum_{m=1}^{p} V_{ms} \sin\frac{m\pi r}{h}. \quad (4.20)$$

4.2.3. B beams with clamped ends under loading on the A beams symmetrical about centre-line xx

For grillages with the B beams clamped at their ends, the procedure differs from that for simple supports and is more intricate. Analysis is only possible for loadings symmetrical about xx. For the mathematical derivations, the reader should refer to the original papers given in the Bibliography.

1st Step: calculation of flexibility coefficients for the A beams. This step is identical to that described in section 4.2.1.

2nd Step: formulation and solution of simultaneous equations in Z coefficients. The general equation is

$$\alpha_s K_{2m} Z_{ms} + \sum_t \beta_{st} Z_{mt} = -\frac{b^3}{8EI_s} \sum_{r=1}^{p} \cos\frac{2m\pi r}{h} \sum_u \beta_{su} P_{ru} + \frac{1}{h} \sum_t \beta_{st} U_t, \quad (4.21)$$

where

$$U_t = \sum_{m=1}^{p} Z_{mt}$$

and the ranges of the summations in t and u are given in table 2. An equation should be written down, using the numerical values of γ_{su} and β_{st} obtained in the first step, for each integer value of s over the ranges given for the summations in t.

Algebraic solution of the set of simultaneous equations for Z_{m_1}, Z_{m_2}, etc., in terms of P_{ru} and U_t then leads to expressions of the form

$$Z_{ms} = f(m, s, P) + \sum_t c_{st} U_t. \quad (4.22)$$

[30]

Next, the sums U_1, U_2, etc., are formed from (4.22) by summation from $m = 1$ to p, giving

$$U_s = \sum_{m=1}^{p} f(m, s, P) + \sum_{m=1}^{p} \sum_t c_{st} U_t. \tag{4.23}$$

Equations (4.23) for $s = 1, 2$, etc., may now be solved for U_1, U_2, etc., and the solutions substituted into equations (4.22), to obtain Z_{m_1}, Z_{m_2}, etc., explicitly. The details of this process will be illustrated later in a numerical example. It will be noticed that the number of simultaneous equations to be solved on each occasion is equal to q, the number of B beams (or half this number in cases of symmetry). Numerical values of Z_{ms} are required for integral values of m up to $\frac{1}{2}(p+1)$.

3rd Step: calculation of deflections, reactions and bending moments.
At the intersection points

$$w_{rs} = \frac{1}{6h^4} \left[2 \sum_{m=1}^{\frac{1}{2}h-1} \left(1 - \cos\frac{2m\pi r}{h}\right) K_{2m} Z_{ms} + (1 - \cos r\pi) K_h Z_{\frac{1}{2}h,\, s} \right], \tag{4.24}$$

$$R_{rs} = -\frac{8EI_s}{hb^3} \left[2 \sum_{m=1}^{\frac{1}{2}h-1} \left(\cos\frac{2m\pi r}{h}\right) Z_{ms} + (\cos r\pi) Z_{\frac{1}{2}h,\, s} \right], \tag{4.25}$$

and the bending moment of the B beams is

$$M_{rs} = \frac{2EI_s}{h^2 b^2} \left[2 \sum_{m=1}^{\frac{1}{2}h-1} \left(\cos\frac{2m\pi r}{h}\, \text{cosec}^2\frac{m\pi}{h}\right) Z_{ms} + (\cos r\pi) Z_{\frac{1}{2}h,\, s} \right]. \tag{4.26}$$

For grillages containing an even number of A beams (p even), the last term of equations (4.24) to (4.26) should be taken as zero. Knowing the reactions, the bending moments of the A beams may be found by statics.

4.3. Formula for simply supported grillages

An explicit and exact formula has been derived for grillages where both the A and B beams are uniform, equal and evenly spaced. For a concentrated load P_{tu} applied at the intersection point t_u, the intersection reactions are given by the formula

$$R_{rs} = \frac{4P_{tu}}{(p+1)(q+1)} \sum_{m=1}^{p} \sum_{n=1}^{q} \frac{\sin\dfrac{m\pi r}{p+1}\sin\dfrac{m\pi t}{p+1}\sin\dfrac{n\pi s}{q+1}\sin\dfrac{n\pi u}{q+1}}{\left(1 + \lambda\dfrac{K_m}{K_n}\right)}, \tag{4.27}$$

[31]

where
$$\lambda = \frac{(q+1)^3 b^3 I_a}{(p+1)^3 a^3 I_b},$$

and the bending moments follow by statics.

4.4.1. Example 1—3×9 beam grillage with simply supported edges under central concentrated load

The grillage to be discussed, illustrated in fig. 4, was previously analysed by various approximate methods. The stiffeners will be considered as the A beams and the girders as the B beams, i.e. $p = 9$, $q = 3$, since this gives the smaller number, $\frac{1}{2}(q+1)$, of simultaneous equations to be solved. Thus we have $a/b = \frac{1}{3}$ and $I_a/I_b = \frac{1}{4}$, where I_b denotes the moment of inertia of the B beams (I_s in the general analysis). The externally applied load P will be considered as acting on the A beam at the centre.

1st Step: calculation of flexibility coefficients for the A beams. The symmetry and loading correspond to case (ii), table 2, and substitution of $x_1/a = \frac{1}{4}$ and $x_2/a = \frac{1}{2}$ into the formulae of table 1 leads to the following values:

$$\beta_{11} = \frac{8}{6 \times 4^3}, \quad \beta_{12} = \frac{11}{3 \times 4^4}, \quad \beta_{21} = \frac{11}{6 \times 4^3}, \quad \beta_{22} = \frac{8}{6 \times 4^3}.$$

We may now write down, by equation (4.14), the following expressions for the deflection of the rth beam:

$$-6 \times 4^3 w_{r1}/B_a = 8R_{r1} + \tfrac{11}{2}(R_{r2} - \alpha_r P),$$

$$-6 \times 4^3 w_{r2}/B_a = 11R_{r1} + 8(R_{r2} - \alpha_r P),$$

where $\alpha_r = 0$, $r \neq 5$ and $\alpha_5 = 1$, $B_a = a^3/EI_a$.

2nd Step: formulation and solution of simultaneous equations in Z coefficients. Since all the B beams are identical, we may write $\alpha_s = \alpha$ and $I_s = I_b$. Then equation (4.15) gives

$$\left. \begin{aligned} 3 \times 4^4 \alpha K_m Z_{m1} + (16Z_{m1} + 11Z_{m2}) &= \frac{22b^3 P}{EI_b} \sin \frac{m\pi}{2}, \\[2mm] 3 \times 4^4 \alpha K_m Z_{m2} + (22Z_{m1} + 16Z_{m2}) &= \frac{32b^3 P}{EI_b} \sin \frac{m\pi}{2}. \end{aligned} \right\} \qquad (4.28)$$

We will work out the analysis for $Pb^3/EI_b = 0$. Now writing

$$3 \times 4^4 \alpha K_m = H_m \quad \text{and} \quad \frac{b^3 P}{EI_b} \sin \frac{m\pi}{2} = \sin \frac{m\pi}{2} = P_m,$$

equations (4.28) become

$$(16+H_m)\,Z_{m1}+11Z_{m2} = 22P_m,$$

$$22Z_{m1}+(16+H_m)\,Z_{m2} = 32P_m,$$

and solving for Z_{m1}, Z_{m2}, we obtain

$$\left.\begin{array}{l} Z_{m1}/P_m = 22H_m/\Delta_m, \\[4pt] Z_{m2}/P_m = (28+32H_m)/\Delta_m, \end{array}\right\} \tag{4.29}$$

where $\Delta_m = 14+32H_m+H_m^2$.

The numerical calculation of the required Z_{ms} coefficients is shown in table 3. It should be noted that for m even, $P_m = Z_{ms} = 0$ (since the loading is symmetrical about xx, fig. 6).

TABLE 3. *Calculation of Z_{m1} and Z_{m2} for example* 1

$(3 \times 4^4\alpha = 16b^3 I_a/(p+1)^3, \quad a^3 I_b = 0{\cdot}108, \quad Pb^3/EI_b = 1.)$

m ...	1	3	5	7	9
$K_m = \mathrm{cosec}^2\dfrac{m\pi}{20}\left(1+3\cot^2\dfrac{m\pi}{20}\right)$	4928	60·92	8	2·241	1·102
$H_m = 3 \times 4^4\alpha K_m$	532·2	6·579	0·864	0·2420	0·1190
$\Delta_m = 14+32H_m+H_m^2$	$3{\cdot}003 \times 10^5$	267·8	42·39	21·80	17·82
P_m	1	−1	1	−1	1
$Z_{m1} \times 10^3$	38·88	−540·4	448·4	−244·2	146·9
$Z_{m2} \times 10^3$	56·81	−890·6	1,312·7	−1,639·5	1·784·7

3rd Step: calculation of deflections, reactions and bending moments. These follow, simply, using equations (4.11), (4.7) and (4.12). Values for positions around the centre of the grillage are shown in table 4. When comparing these answers with those obtained in chapter 3, the symbols a, b, I_b become b, a, I_g respectively.

TABLE 4. *Deflection, bending moment and shear force results for example* 1

Quantity		
Deflections of central stiffener	$w_{51} \times 10^5 EI_b/Pb^3$	47·78
	$w_{52} \times 10^5 EI_b/Pb^3$	72·99
Bending moment at centre	Girder $M_{52} \times 10^3/Pb$	32·91
	Stiffener $M'_{52} \times 10^3/Pb$	24·13
Shear forces over spans adjoining centre	Girder $F \times 10^2/P$	28·42
	Stiffener $F' \times 10^2/P$	21·58

4.4.2. Example 2—3×9 beam grillage with clamped edges under central concentrated load

The analysis of example 1 will now be repeated for the ends of the beams clamped against rotation.

1st Step: calculation of flexibility coefficients for the A beams. The nine stiffeners will again be considered as the A beams, and the formulae of table 1 for clamped ends lead to the following influence coefficients for these beams:

$$\beta_{11} = \frac{5}{6 \times 4^4}, \quad \beta_{12} = \frac{1}{6 \times 4^3}, \quad \beta_{21} = \frac{1}{3 \times 4^3} \quad \text{and} \quad \beta_{22} = \frac{1}{3 \times 4^3}.$$

We may now write down, by equation (4.14), the following expressions for the deflection of the rth A beam:

$$-6 \times 4^4 w_{r1}/B_a = 5R_{r1} + 4(R_{r2} - \alpha_r P),$$

$$-6 \times 4^4 w_{r2}/B_a = 8R_{r1} + 8(R_{r2} - \alpha_r P),$$

where $\alpha_r = 0, r \neq 5$ and $\alpha_5 = 1$.

2nd Step: formulation and solution of simultaneous equations in Z coefficients. Writing $\alpha_s = \alpha$ (all B beams identical), equation (4.21) gives

$$6 \times 4^4 \alpha K_{2m} Z_{m1} + (5Z_{m1} + 4Z_{m2}) = -\frac{b^3}{2EI_b} (\cos m\pi) P + (5U_1 + 4U_2)/10,$$
$$6 \times 4^4 \alpha K_{2m} Z_{m2} + (8Z_{m1} + 8Z_{m2}) = -\frac{b^3}{EI_b} (\cos m\pi) P + (8U_1 + 8U_2)/10. \quad \left.\right\}$$

$$(4.30)$$

We will work out the analysis for $Pb^3/2EI_b = 1$. Now, writing

$$6 \times 4^4 \alpha K_{2m} = L_m \quad \text{and} \quad b^3(\cos m\pi) P/2EI_b = \cos m\pi = P_m,$$

equations (4.30) become

$$(5 + L_m) Z_{m1} + 4Z_{m2} = -P_m + (5U_1 + 4U_2)/10,$$

$$8Z_{m1} + (8 + L_m) Z_{m2} = -2P_m + (8U_1 + 8U_2)/10,$$

and solving for Z_{m1}, Z_{m2} in terms of the quantities on the right, we obtain

$$\Delta_m Z_{m1} = -L_m P_m + [(8 + 5L_m) U_1 + 4L_m U_2]/10,$$
$$\Delta_m Z_{m2} = -2(1 + L_m) P_m + [8L_m U_1 + 8(1 + L_m) U_2]/10, \quad \left.\right\} \quad (4.31)$$

where $\Delta_m = 8 + 13L_m + L_m^2$.

[34]

Next, forming the sums U_1, U_2 from (4.31) by multiplying by Δ_m^{-1} and taking the sums over $m = 1, 2, ..., 9$.

$$U_1 = \sum_{m=1}^{9} \Delta_m^{-1}(-L_m P_m + [(8 + 5L_m)\, U_1 + 4L_m U_2]/10),$$

$$U_2 = \sum_{m=1}^{9} \Delta_m^{-1}(-2(1 + L_m)\, P_m + [8L_m U_1 + 8(1 + L_m)\, U_2]/10).$$

By the substitutions,

$$\left. \begin{aligned}
A_1 &= \sum_{m=1}^{9} \Delta_m^{-1} = 2 \sum_{m=1}^{4} \Delta_m^{-1} + \Delta_5^{-1}, \\
A_2 &= \sum_{m=1}^{9} L_m \Delta_m^{-1} = 2 \sum_{m=1}^{4} L_m \Delta_m^{-1} + L_5 \Delta_5^{-1}, \\
A_3 &= \sum_{m=1}^{9} P_m \Delta_m^{-1} = 2 \sum_{m=1}^{4} P_m \Delta_m^{-1} + P_5 \Delta_5^{-1}, \\
A_4 &= \sum_{m=1}^{9} L_m P_m \Delta_m^{-1} = 2 \sum_{m=1}^{4} L_m P_m \Delta_m^{-1} + L_5 P_5 \Delta_5^{-1},
\end{aligned} \right\} \quad (4.32)$$

the above equations can be written in the form

$$\left. \begin{aligned}
U_1(1 - 0\cdot 8A_1 - 0\cdot 5A_2) - 0\cdot 4A_2 U_2 &= -A_4, \\
-0\cdot 8A_2 U_1 + U_2(1 - 0\cdot 8A_1 - 0\cdot 8A_2) &= -2(A_3 + A_4).
\end{aligned} \right\} \quad (4.33)$$

The solution of equations (4.33) is

$$\left. \begin{aligned}
AU_1 &= -0\cdot 8A_2 A_3 - (1 - 0\cdot 8A_1)\, A_4, \\
AU_2 &= -2(1 - 0\cdot 8A_1 - 0\cdot 5A_2)\, A_3 - 2(1 - 0\cdot 8A_1 - 0\cdot 1A_2)\, A_4,
\end{aligned} \right\} \quad (4.34)$$

where $A = 1 - 1\cdot 6A_1 - 1\cdot 3A_2 + 0\cdot 64A_1^2 + 1\cdot 04A_1 A_2 + 0\cdot 08A_2^2$.

Thus, to calculate the coefficients, Z_{ms}, it is necessary to proceed as follows (see table 5):

(a) Calculate L_m, Δ_m and P_m for $m = 1, 2, ..., 5$, and hence the sums A_1, A_2, A_3, A_4 by equations (4.32).

(b) Evaluate U_1, U_2 by equations (4.34),

$$A = 0\cdot 2288, \quad U_1 = -0\cdot 07239, \quad U_2 = -0\cdot 12648.$$

(c) Substitute U_1, U_2 into equations (4.31), to obtain Z_{m1}, Z_{m2}, explicitly. As an example,

$$Z_{51} = [-L_5 P_5 + (0\cdot 8 + 0\cdot 5L_5)\, U_1 + 0\cdot 4L_5 U_2]/\Delta_5 = 0\cdot 01284.$$

The complete list of Z_{ms} coefficients required is shown in table 6.

TABLE 5. *Calculation of A_1, A_2, A_3 and A_4 for example 2*

$(6 \times 4^4 \alpha = 32 b^3 I_a / (p+1)^3 b^3 I_b = 0 \cdot 216, \quad Pb^3/2EI_b = 1.)$

m ...	1	2	3	4	5	$\sum\limits_{m=1}^{9}$
$K_{2m} = \operatorname{cosec}^2\left(\dfrac{m\pi}{10}\right)$ $\times \left[1 + 3\cot^2\left(\dfrac{m\pi}{10}\right)\right]$	308·1	19·34	3·947	1·456	1·0	—
$L_m = 6 \times 4^4 \alpha K_{2m}$	66·54	4·178	0·8526	0·3144	0·216	—
$\Delta_m = 8 + 13 L_m + L_m^2$	5301	79·78	19·81	12·187	10·855	—
$P_m = \cos m\pi$	−1	1	−1	1	−1	—
Δ_m^{-1}	0·00019	0·01253	0·05048	0·08206	0·09212	$A_1 =$ 0·3826
$L_m \Delta_m^{-1}$	0·01255	0·05238	0·04304	0·02580	0·01990	$A_2 =$ 0·2874
$P_m \Delta_m^{-1}$	−0·00019	0·01253	−0·05048	0·08206	−0·09212	$A_3 =$ 0·00428
$L_m P_m \Delta_m^{-1}$	−0·01255	0·05238	−0·04304	0·02580	−0·01990	$A_4 =$ 0·02528

TABLE 6. *Values of Z_{ms} for example 2*

$(Pb^3/2EI_b = 1.)$

	$s = 1$	$s = 2$
$m = 1$	0·01145	0·02347
$m = 2$	−0·05765	−0·1394
$m = 3$	0·03638	0·1751
$m = 4$	−0·03279	−0·2281
$m = 5$	0·01284	0·2116

3rd Step: calculation of deflections, reactions and bending moments. These follow simply, using equations (4.24), (4.25) and (4.26). Values of the deflection, bending moment and shear force around the centre of the grillage are given in table 7.

In general, no significant figures are lost in a grillage calculation, though figures may be lost in some cases. Thus, to obtain a solution accurate to within 1 %, computation by slide rule should be sufficiently accurate, but a careful watch should be kept for any loss of significant figures.

[36]

It should be noted that, although the maximum bending moment often occurs at the loading position, this is not true in general, and it is usually necessary to work out other bending moments to determine the maximum value with certainty.

TABLE 7. *Deflection, bending moment and shear force results for example 2*

Quantity		
Deflections of central stiffener	$w_{51} \times 10^5 EI_b/Pb^3$	12·26
	$w_{52} \times 10^5 EI_b/Pb^3$	26·74
Bending moments at centre	Girder $M_{52} \times 10^3/Pb$	25·54
	Stiffener $M'_{52} \times 10^3/Pb$	16·84
Shear forces over spans adjoining	Girder $F \times 10^2/P$	26·91
centre	Stiffener $F' \times 10^2/P$	23·09

It may be seen that analysis by the method of Fourier Series is more lengthy and troublesome for clamped edges, than for simple supports.

4.4.3. Example 3—3 × 3 beam simply supported grillage under off-centre concentrated load

For grillages with an intersection at the centre, the maximum bending moment under a single concentrated load only seldom occurs when the load is applied at the centre. More often, larger bending moments arise when the load is applied between intersections, as in the grillage of fig. 7. This very simple square grillage is identical in the two orthogonal directions, with three equal, uniform and evenly spaced beams in each direction. We will analyse this grillage with simply supported edges, for a load applied between beam intersection points, by two methods: first, by considering the loaded beam as a *B* beam (to illustrate the analysis for loading along a *B* beam), and secondly, considering the loaded beam as an *A* beam (to illustrate the treatment of unsymmetrical loading by superposition).

(a) *Loaded beam considered as a B beam.* The formulae of table 1 for simple supports lead to the following influence coefficients for the *A* beams:

$$\beta_{11} = \frac{8}{6 \times 4^3}, \quad \beta_{12} = \frac{11}{3 \times 4^4}, \quad \beta_{21} = \frac{11}{6 \times 4^3} \quad \text{and} \quad \beta_{22} = \frac{8}{6 \times 4^3}.$$

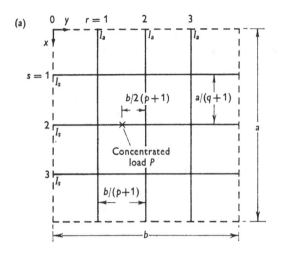

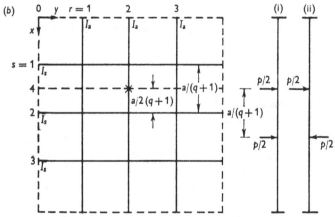

External loading considered as sum of
symmetrical load (i) and anti-symmetrical load (ii)

Fig. 7. 3×3 grillage identical in two orthogonal directions. (a) Loaded beam considered as B beam. (b) Loaded beam considered as A beam.

Then, writing $\alpha_s = \alpha$ (all B beams identical), equation (4.17) gives

$$
\left.
\begin{aligned}
3 \times 4^4 \alpha (K_m V_{m1}) &= -16 V_{m1} - 11 V_{m2}, \\
3 \times 4^4 \alpha (K_m V_{m2} + J_m P \sin \tfrac{3}{8} m\pi) &= -22 V_{m1} - 16 V_{m2}.
\end{aligned}
\right\} \quad (4.35)
$$

We will work out the analysis for $P = 1$. Now, writing $3 \times 4^4 \alpha K_m = H_m$ and $3 \times 4^4 \alpha J_m P \sin \tfrac{3}{8} m\pi = P_m$, equations (4.35) become

$$
(16 + H_m) V_{m1} + 11 V_{m2} = 0, \quad 22 V_{m1} + (16 + H_m) V_{m2} = -P_m,
$$

and, solving for V_{m1}, V_{m2}, we obtain

$$V_{m1} = 11P_m/\Delta_m, \quad V_{m2} = -(16+H_m)\,P_m/\Delta_m,$$

where $\Delta_m = 14 + 32H_m + H_m^2.$

The numerical values of the V_{ms} coefficients required, and the bending moment at the loaded point, may now be calculated. The latter is obtained from the reactions R_{12} and R_{22} (see equation 4.20 and table 8) by statics,

$$\frac{M}{Pb} = \frac{15}{64} + \frac{5}{32}\frac{R_{12}}{P} + \frac{3}{16}\frac{R_{22}}{P} + \frac{3}{32}\frac{R_{32}}{P}, \tag{4.36}$$

and $M/Pb = 0{\cdot}0996$, by equation (4.34).

TABLE 8. *Calculation of reactions for example* 3 (*a*)

m ...	1	2	3	$\dfrac{1}{4}\displaystyle\sum_{m=1}^{3}$
K_m	126·2	8·0	1·780	—
J_m	252·2	15·56	2·718	—
$H_m = 3 \times 4^4 \alpha K_m$	31·56	2·0	0·445	—
$P_m = 3 \times 4^4 J_m \sin \tfrac{3}{8}m\pi$	58·25	2·75	−0·2597	—
$\Delta_m = 14 + 32H_m + H_m^2$	2020	82	28·44	—
$V_{m2} = -(16+H_m)\,P_m/\Delta_m$	−1·372	−0·604	0·1502	—
$V_{m2} \sin \tfrac{1}{4}m\pi$	−0·970	−0·604	0·106	$R_{12} = -0{\cdot}367$
$V_{m2} \sin \tfrac{1}{2}m\pi$	−1·372	0	−0·150	$R_{22} = -0{\cdot}381$
$V_{m2} \sin \tfrac{3}{4}m\pi$	−0·970	0·604	0·106	$R_{32} = -0{\cdot}065$

(*b*) *Loaded beam considered as an A beam.* As already stated, for unsymmetrical loading applied to a grillage having symmetry in the x direction, the quickest solution is obtained by superposition of a symmetrical and an anti-symmetrical load, as in figs. 3, 6 and 7(*b*). Considering, first, the symmetrical loading (i), the formulae of table 1 give the following influence coefficients for the A beams:

$$\beta_{11} = \frac{8}{6\times 4^3}, \quad \beta_{12} = \frac{11}{3\times 4^4}, \quad \beta_{21} = \frac{11}{6\times 4^3} \quad \text{and} \quad \beta_{22} = \frac{8}{6\times 4^3};$$

$$\beta_{14} = \frac{20{\cdot}5}{3\times 4^4} \quad \text{and} \quad \beta_{24} = \frac{29{\cdot}25}{3\times 4^4}.$$

[39]

For $Pb^3/EI_s = 1$, equation (4.15) gives

$$(16 + H_m) Z_{m1} + 11 Z_{m2} = 20 \cdot 5 P_m,$$

$$22 Z_{m1} + (16 + H_m) Z_{m2} = 29 \cdot 25 P_m,$$

where $H_m = 3 \times 4^4 \alpha K_m$, $P_m = \sin \frac{1}{2} m \pi$, and solving for Z_{m1}, Z_{m2}, we obtain

$$Z_{m1} = (6 \cdot 25 + 20 \cdot 5 H_m) P_m / \Delta_m, \quad Z_{m2} = (17 + 29 \cdot 25 H_m) P_m / \Delta_m,$$

where $\Delta_m = 14 + 32 H_m + H^2$.

The numerical values of the Z_{ms} coefficients and the reactions R_{21} and R_{22} (equation (4.7)) may now be worked out (table 9).

TABLE 9. *Calculation of reactions for example 3(b), loading (i)*

m	1	3	$\dfrac{P}{4} \overset{3}{\underset{m=1}{\Sigma}}$
K_m	126·2	1·780	—
$H_m = 3 \times 4^4 \alpha K_m$	31·56	0·445	—
$\Delta_m = 14 + 32 H_m + H_m^2$	2020	28·39	—
$Z_{m1}/P_m = (6 \cdot 25 + 20 \cdot 5 H_m)/\Delta_m$	0·3234	0·5405	$R_{21} = 0 \cdot 2160 P$
$Z_{m2}/P_m = (17 + 21 \cdot 25 H_m)/\Delta_m$	0·4654	1·0557	$R_{22} = 0 \cdot 3802 P$

(For this loading, symmetrical about xx, fig. 6, $Z_{ms} = 0$ for m even.)

Considering now, the anti-symmetrical loading (ii), the relevant influence coefficient for the A beams (from table 1) is

$$\beta_{11} = \frac{1}{6 \times 4^3}.$$

For $Pb^3/EI_s = 1$, equation (4.15) gives

$$(2 + H_m) Z_{m1} = 1 \cdot 375 P_m.$$

Numerical calculation of Z_{m1} and the reaction follows simply (table 10).

TABLE 10. *Calculation of R_{21} for example 3(b), loading (ii)*

	$(3 \times 4^4 \alpha = 0 \cdot 25, \quad Pb^3/EI_s = 1)$		
m	1	3	$\dfrac{P}{4} \overset{3}{\underset{m=1}{\Sigma}}$
$Z_{m1}/P_m = 1 \cdot 375/(2 + H_m)$	0·0410	0·5627	$R_{21} = 0 \cdot 1510 P$

For this anti-symmetrical loading, $R_{22} = 0$.

Now, superimposing loadings (i) and (ii) (fig. 7), we obtain

$$R_{21} = 0 \cdot 3670P, \quad R_{22} = 0 \cdot 3802P \quad \text{and} \quad R_{23} = 0 \cdot 0650P.$$

The bending moment at the loaded point may be obtained by statics,

$$\frac{M}{Pa} = \frac{15}{64} - \frac{5}{32}\frac{R_{21}}{P} - \frac{3}{16}\frac{R_{22}}{P} - \frac{3}{32}\frac{R_{23}}{P} = 0 \cdot 0997,$$

agreeing with the value obtained by method (a). It is interesting to note that for central loading of this grillage, the maximum bending moment (at the centre) is $22\frac{1}{2}\%$ lower,

$$M/Pa = 0 \cdot 0820.$$

Chapter 5

SOLUTION BY DISPLACEMENT METHOD USING ELECTRONIC COMPUTER

In earlier chapters, several methods have been described which avoid solving the large sets of simultaneous equations obtained in a straight-forward exact solution, Fortunately, it is no longer necessary to resort to these approximate or intricate methods, since large sets of simultaneous equations can be readily solved using a digital electronic computer, and present-day medium-sized computers can tackle up to about seventy equations; the large computers can solve consider-ably more. An important extension of the use of computers is to arrange for the entire calculation to be carried out on the computer, including formulation of the equations, their solution, and evaluation of the deflections and stresses. Once such a program of instructions has been devised, a grillage calculation merely consists of specifying the geometry and sizes of the members of the grillage in an ordered sequence, and interpreting the tables of deflections and stresses which are printed out at the conclusion of the calculation.

In this chapter, a general numerical solution by the displacement method is described. Straight segments of beams between the inter-section points are analysed in terms of the deflection and two ortho-gonal slopes at each intersection. The forces and moments acting at the ends of the beam elements between the intersections are calcu-lated in terms of these deflections and slopes, and the condition that the total forces and moments at each intersection are in equilibrium provides a set of simultaneous equations in the displacements. This solution formulated in the displacements is the inverse treatment to the formulation in forces described in chapter 2. In the form described here, it involves three unknowns at each intersection instead of one by the Force Method, but the advantage is that torsional rigidity can be included, and any shape of boundary or arrangement of beams can be treated, including non-orthogonal plan forms. The terms due to shear deflection are also included. A computer program, based on this method, has been developed for the Ferranti Pegasus com-puter, and the logic of the method which is outlined in this chapter can be readily adopted for other computers. The numerical effects of shear and torsion are discussed in chapter 6.

[42]

5.1. Analysis of a beam element including shear and torsion

A straight segment or element of a beam under a general set of forces and moments applied at each end, and loaded laterally between the ends, will be analysed by simple beam theory including shear deflections, to determine the total deflection, the bending slope and the angle of twist at each end. Only the bending term of the slope is considered since the shear slopes do not have to be compatible in the grillage calculation. A common example of this effect is at the ends of a clamped beam where, although bending slopes are prevented, the beam is distorted giving a shear slope immediately inwards from the clamps. The equations relating the deflections and rotations to the forces and moments are inverted to obtain explicit expressions for the forces and moments.

Considering first the action under the bending and shear forces, using the notation given in fig. 8, we have the following differential equations:

$$EI \frac{d^2 w_b}{dx^2} = -M_0 + S_0 x + W(x - \alpha B) U(x - \alpha B) + \tfrac{1}{2} p x^2, \quad (5.1)$$

where $\qquad U(x - \alpha B) = 0 \quad \text{for} \quad (x - \alpha B) \text{ negative}$

$$= 1 \quad \text{for} \quad (x - \alpha B) \text{ positive},$$

$$GA \frac{d w_s}{dx} = -S_0 - W U(x - \alpha B) - p x. \quad (5.2)$$

Solution of equations (5.1) and (5.2) gives the following expressions for the bending slope dw_b/dx and the total deflection:

$$\frac{dw_b}{dx} = v_0 - \frac{M_0 x}{EI} + \frac{S_0 x^2}{2EI} + \frac{W(x - \alpha B)^2 U(x - \alpha B)}{2EI} + \frac{p x^3}{6EI},$$

$$w = w_0 + v_0 x - \frac{M_0 x^2}{2EI} + \frac{S_0 x^3}{6EI} + \frac{W(x - \alpha B)^3 U(x - \alpha B)}{6EI} + \frac{p x^4}{24EI}$$

$$- \frac{S_0 x}{GA} - \frac{W(x - \alpha B) U(x - \alpha B)}{GA} - \frac{p x^2}{2GA}.$$

Substituting $x = B$, expressions are obtained for v_B and w_B, which may be solved to give M_0 and S_0 explicitly:

$$M_0 = + \frac{4EI(1 + 3\beta)}{B(1 + 12\beta)} v_0 + \frac{6EI}{B^2(1 + 12\beta)} w_0 + \frac{2(1 - 6\beta) EI}{B(1 + 12\beta)} v_B$$

$$- \frac{6EI}{B^2(1 + 12\beta)} w_B - \frac{W B(1 - \alpha) \alpha (1 - \alpha + 6\beta)}{1 + 12\beta} - \tfrac{1}{12} p B^2, \quad (5.3)$$

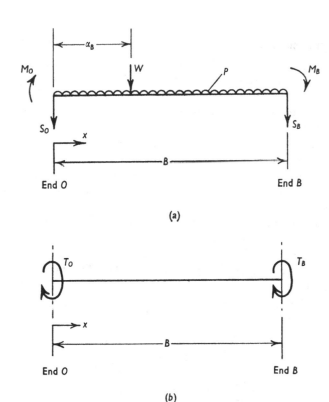

(a)

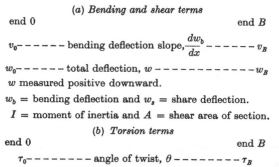

(b)

Fig. 8. Notation for analysis of beam element.

(a) *Bending and shear terms*

end 0   end B

v_0 – – – – – bending deflection slope, $\dfrac{dw_b}{dx}$ – – – – – – v_B

w_0 – – – – – total deflection, w – – – – – – – – – – – w_B

w measured positive downward.

w_b = bending deflection and w_s = shear deflection.

I = moment of inertia and A = shear area of section.

(b) *Torsion terms*

end 0   end B

τ_0 – – – – – – – angle of twist, θ – – – – – – – τ_B

T_0 and T_B are positive for clockwise moments when viewed from end O.

τ_0 and τ_B are positive for clockwise slopes relative to horizontal plane through beam, viewed from end O.

J = equivalent polar moment of inertia for torsion of section.

[44]

$$S_0 = \frac{6EI}{B^2(1+12\beta)}v_0 + \frac{12EI}{B^3(1+12\beta)}w_0 + \frac{6EI}{B^2(1+12\beta)}v_B$$

$$- \frac{12EI}{B^3(1+12\beta)}w_B - \frac{W(1-\alpha)\left[(1-\alpha)(1+2\alpha)+12\beta\right]}{1+12\beta} - \tfrac{1}{2}pB, \quad (5.4)$$

where $\beta = EI/B^2GA$.

The values of M_B and S_B are obtained by statics:

$$M_B = -M_0 + S_0 B + WB(1-\alpha) + \tfrac{1}{2}pB^2, \tag{5.5}$$

$$S_B = -S_0 - W - pB. \tag{5.6}$$

Equations (5.3)–(5.6) determine the forces and bending moments at the ends of the beam element in terms of the total deflection and bending slope at these ends and the loading externally applied between the ends. It should be noted that the shear area coefficient β enters into all the terms of these expressions for the end forces with the exception of the symmetrical uniform load term in p. The inclusion of shear deflection is not simply a question of adding shear deflections to the complete solution from the bending analysis.

We will consider now the action under the torques T_0 and T_B. The differential equation is

$$GJ\frac{d\theta}{dx} = T_B,$$

and by statics $T_0 = -T_B$. Simple analysis leads to the following solution:

$$T_0 = +\frac{GJ}{B}\tau_0 - \frac{GJ}{B}\tau_B, \tag{5.7}$$

$$T_B = -\frac{GJ}{B}\tau_0 + \frac{GJ}{B}\tau_B. \tag{5.8}$$

Equations (5.3)–(5.8) provide a complete solution for the forces, bending and twisting moments at the ends of the beam element, and they may be rewritten in the form of matrix equations as follows:

$$\begin{bmatrix} M_0 \\ T_0 \\ S_0 \end{bmatrix} = \begin{bmatrix} a_{11} & a_{12} & a_{13} \\ a_{21} & a_{22} & a_{23} \\ a_{31} & a_{32} & a_{33} \end{bmatrix} \begin{bmatrix} v_0 \\ \tau_0 \\ w_0 \end{bmatrix} + \begin{bmatrix} a_{14} & a_{15} & a_{16} \\ a_{24} & a_{25} & a_{26} \\ a_{34} & a_{35} & a_{36} \end{bmatrix} \begin{bmatrix} v_B \\ \tau_B \\ w_B \end{bmatrix} + \begin{bmatrix} a_{17} \\ a_{27} \\ a_{37} \end{bmatrix},$$

$$\tag{5.9}$$

$$\begin{bmatrix} M_B \\ T_B \\ S_B \end{bmatrix} = \begin{bmatrix} b_{11} & b_{12} & b_{13} \\ b_{21} & b_{22} & b_{23} \\ b_{31} & b_{32} & b_{33} \end{bmatrix} \begin{bmatrix} v_B \\ \tau_B \\ w_B \end{bmatrix} + \begin{bmatrix} b_{14} & b_{15} & b_{16} \\ b_{24} & b_{25} & b_{26} \\ b_{34} & b_{35} & b_{36} \end{bmatrix} \begin{bmatrix} v_0 \\ \tau_0 \\ w_0 \end{bmatrix} + \begin{bmatrix} b_{17} \\ b_{27} \\ b_{37} \end{bmatrix},$$

$$\tag{5.10}$$

where

$$a_{11} = b_{11} = EI(4 + 12\beta)/B(1 + 12\beta),$$

$$a_{12} = b_{12} = 0,$$

$$a_{13} = -b_{13} = 6EI/B^2(1 + 12\beta),$$

$$a_{21} = b_{21} = 0,$$

$$a_{22} = b_{22} = GJ/B,$$

$$a_{23} = b_{23} = 0,$$

$$a_{31} = -b_{31} = a_{13},$$

$$a_{32} = b_{32} = 0,$$

$$a_{33} = b_{33} = 12EI/B^3(1 + 12\beta),$$

$$a_{14} = b_{14} = 2EI(1 - 6\beta)/B(1 + 12\beta),$$

$$a_{15} = b_{15} = 0,$$

$$a_{16} = -b_{16} = -a_{13},$$

$$a_{24} = b_{24} = 0,$$

$$a_{25} = b_{25} = -a_{22},$$

$$a_{26} = b_{26} = 0,$$

$$a_{34} = -b_{34} = a_{13},$$

$$a_{35} = b_{35} = 0,$$

$$a_{36} = b_{36} = -a_{33},$$

$$a_{17} = -(\tfrac{1}{12}pB^2) - WB(1 - \alpha)\alpha[(1 - \alpha) + 6\beta]/(1 + 12\beta),$$

$$b_{17} = (\tfrac{1}{12}pB^2) + WB(1 - \alpha)\alpha[\alpha + 6\beta]/(1 + 12\beta),$$

$$a_{27} = b_{27} = 0,$$

$$a_{37} = -(\tfrac{1}{2}pB) - W(1 - \alpha)[(1 + 2\alpha)(1 - \alpha) + 12B]/(1 + 12\beta),$$

$$b_{37} = -(\tfrac{1}{2}pB) - W\alpha[\alpha(3 - 2\alpha) + 12\beta]/(1 + 12\beta).$$

5.2. Outline of grillage solution

To analyse the complete grillage, it is convenient to refer to the lengths of beam between intersections as 'elements', and to the intersections as 'hubs'. In order to take full account of symmetry according to the scheme indicated in fig. 3, and to allow for the particular boundary conditions for an element adjacent to the edges of the grillage, it is necessary to consider several different types of element, and these are of two main classes, namely:

(a) Double-ended elements—between two intersections with or without an axis of symmetry or asymmetry at one end.

[46]

(*b*) Single-ended elements—between an intersection and a boundary, or between two intersections with an axis of symmetry or asymmetry intersecting the mid-point.

Double-ended elements are characterized by the need to know the displacements at both ends to determine uniquely the displacements and stresses throughout the element. Single-ended elements are characterized by the need to know the displacements at only one end or point to determine the deflections and stresses throughout the element. The analysis of a general double-ended element has been given above, and equations similar to (5.9) and (5.10) may be obtained for other types of element, the expressions for the elements of the '*a*' and '*b*' matrices being different.

Any number of elements may be linked at each hub. Let θ denote the direction of an element OB, measured positive anticlockwise from the x axis, using a left-handed rectangular co-ordinate system, Oxy. Then, the equations relating the rotations and moments for an axis taken along the beam (given in equations (5.9) and (5.10)) to the corresponding values for the reference axes Oxy are as follows:

$$\left.\begin{array}{ll} \tau = v_x \sin\theta + v_y \cos\theta, & M_x = M\cos\theta + T\sin\theta, \\ v = v_x \cos\theta - v_y \sin\theta, & M_y = -M\sin\theta + T\cos\theta. \end{array}\right\} \quad (5.11)$$

It is then a simple matter to transform equations (5.9) and (5.10) into a similar set in terms of the rotations and moments for the axes Oxy. The general relation for a double-ended element is of the form:

$$\begin{bmatrix} M_{x0} \\ M_{y0} \\ S_0 \end{bmatrix} = \begin{bmatrix} A_{11} & A_{12} & A_{13} \\ A_{21} & A_{22} & A_{23} \\ A_{31} & A_{32} & A_{33} \end{bmatrix} \begin{bmatrix} v_{x0} \\ v_{y0} \\ w_0 \end{bmatrix} + \begin{bmatrix} A_{14} & A_{15} & A_{16} \\ A_{24} & A_{25} & A_{26} \\ A_{34} & A_{35} & A_{36} \end{bmatrix} \begin{bmatrix} v_{xB} \\ v_{yB} \\ w_B \end{bmatrix} + \begin{bmatrix} A_{17} \\ A_{27} \\ A_{37} \end{bmatrix},$$

$$(5.12)$$

$$\begin{bmatrix} M_{xB} \\ M_{yB} \\ S_B \end{bmatrix} = \begin{bmatrix} B_{11} & B_{12} & B_{13} \\ B_{21} & B_{22} & B_{23} \\ B_{31} & B_{32} & B_{33} \end{bmatrix} \begin{bmatrix} v_{xB} \\ v_{yB} \\ w_B \end{bmatrix} + \begin{bmatrix} B_{14} & B_{15} & B_{16} \\ B_{24} & B_{25} & B_{26} \\ B_{34} & B_{35} & B_{36} \end{bmatrix} \begin{bmatrix} v_{x0} \\ v_{y0} \\ w_0 \end{bmatrix} + \begin{bmatrix} B_{17} \\ B_{27} \\ B_{37} \end{bmatrix},$$

$$(5.13)$$

and for a single-ended element,

$$\begin{bmatrix} M_{x0} \\ M_{y0} \\ S_0 \end{bmatrix} = \begin{bmatrix} C_{11} & C_{12} & C_{13} \\ C_{21} & C_{22} & C_{23} \\ C_{31} & C_{32} & C_{33} \end{bmatrix} \begin{bmatrix} v_{x0} \\ v_{y0} \\ w_0 \end{bmatrix} + \begin{bmatrix} C_{14} \\ C_{24} \\ C_{34} \end{bmatrix}. \quad (5.14)$$

Equating to zero the sums of the moments M_x, M_y and the sums of the forces S at every hub provides a sufficient set of equations to find the unknown hub rotations and deflections, v_x, v_y and w. In the

[47]

computer program developed for Pegasus, the evaluation of the elements of the stiffness matrices, A, B and C, the formulation of the equations, and their solution, is done automatically by the computer which then prints out the displacements and evaluates and prints out the moments and forces. The computer also evaluates and prints out the stresses at the ends of every element and at any concentrated load or maximum stress position between the hubs, using the moments and forces with the appropriate beam section moduli.

5.3. Preparation of data tapes for computer

In order to carry out the grillage calculation outlined above, using the computer program, it is first necessary to specify the particular structure and loading to be analysed. This consists of numbering the elements and hubs of the structure in an ordered sequence, specifying how the elements are linked together, and specifying the dimensions, section properties and loading on each element. All this information is listed and punched on to a data tape ready for the computer. The program tape which states how the calculation shall be carried out, together with the data tape, is then fed into the computer which proceeds with the calculation. The preparation of the data tape will now be examined in more detail.

The first step is to number the elements and hubs in the following sequence:

(a) *Double-ended elements.* Every element is numbered at each end using consecutive numbers starting with 1. It is normal, but not necessary, to number the left hand or bottom end of type 1 elements (see below) with an odd number, the other end taking the next consecutive even number. Types 2 and 3 elements are numbered with the even number at the end about which there is symmetry or asymmetry. The last label number for a double-ended element is $2n_1$, where n_1 is the number of these elements. If there are no double-ended elements, the single-ended elements are numbered upwards from 1.

(b) *Single-ended elements.* Every element is numbered at one end starting at $2n_1 + 1$ and ending at $2n_1 + n_2$, where n_2 is the number of single-ended elements. If there are no single-ended elements, the hubs are numbered upwards from $2n_1 + 1$.

(c) *Hubs.* Every hub is given a label starting at $2n_1 + n_2 + 1$ and ending at $2n_1 + n_2 + n_3$, where n_3 is the number of hubs.

With the particular computer being used, there are certain limitations to the size of grillage which can be tackled, and this broadly means that, for one or two independent loading conditions, approximately 20 hubs can be considered, and allowing for symmetry, grillage containing up to about 80 intersections can be analysed.

A fair variety of types of element can be analysed using the program, and the most important are listed below:

	Type number	
Double ended	1	General beam between intersections
	2	Beam with symmetry about one end
	3	Beam with asymmetry about one end
Single ended	11	Beam elastically restrained at end B
	12	Beam simply supported at end B
	13	Beam clamped at end B
	14	Beam symmetrical about mid-point
	15	Beam asymmetrical about mid-point

Information giving the type number, when printed on the data tape, tells the computer which particular subroutine to refer to when evaluating the elements of the stiffness matrices A, B or C in equations (5.12)–(5.14).

The next stage is to prepare the linking data. This consists of the following sequence of numbers.

r	number of independent loading conditions
$2n_1 + 1$	first single-ended element
$2n_1 + n_2 + 1$	first hub
$2n_1 + n_2 + n_3$	last hub
$2n_1 + n_2 + 1$	first hub
d ⎫	pairs of numbers for all double-ended elements at first hub (first element number written down) followed by number
h_c ⎭	of hub to which element is attached
s ⎫	pairs of numbers for all single-ended elements at first hub
o ⎭	together with $+0$

The order in which these numbered pairs is taken is immaterial but all the elements at this hub must be included and when these are exhausted the number -1 is written. The next highest hub number is then taken and the process of numbering repeated, until all the hubs have been taken. Finally, after the last -1 has been written, this part of the data tape ends with

$2n_1 + n_2 + n_3 + 1$	and
L	a warning character

The second portion of the data tape is concerned with the beam data, which are given for each element taken in order of their label numbers. The continuing sequence of numbers is as follows:

N_1 label number of element
N_2 type number

If N_1 is written as a positive number, the following data are given next:

A effective shear area (usually taken as area of web)
I moment of inertia
z distance from neutral axis to outermost fibre, positive if measured to bottom flange and negative if measured to top flange

If N_1 is written as a negative number, the following data are given in place of A, I and z:

b breadth of top flange
h_b thickness of top flange
d depth of web
h_w thickness of web   If this alternative is used, the computer works out A (taken as dh_w), I and z
f breadth of bottom flange
h_f thickness of bottom flange

Then, continue as follows:

B length of element (for types 14 and 15, B is taken as twice the length of the element from the hub to the axis of symmetry or asymmetry)
θ angle element makes with x axis
J equivalent polar moment of inertia for torsion (standard value $+0$)
β twice angle between element and axis of symmetry or angle between element and its image (standard value $+180$)
E Young's modulus (standard value 3×10^7)
G shear modulus (standard value $3 \times 10^7/2 \cdot 6$)
L terminating character
p uniform load/unit length   or p uniform load
$+0$ no concentrated load   , W concentrated load
                                α position of load
or $+0$ no uniform load   or $+0$ no uniform load
     $+0$ no concentrated load W concentrated load
                                    α position of load

A sequence of such terms is given for each loading condition. For type 11 elements only, the following are also given:

A elastic restraint coefficient for moments, $M_B = -Av_B$
B elastic restraint coefficient for torques, $T_B = -B\tau_B$
C elastic restraint coefficient for forces, $S_B = -Cw_B$

The last four parameters before the L (J, β, E and G) need to be included only if they differ from the standard values given. If any of these values are non-standard, the values $J, ..., G$ must be specified whether they be standard or non-standard, up to and including the last non-standard value. Concentrated loads applied to the grillage at an intersection may be considered as acting at either end of an element ($\alpha = 0$ or 1).

In order to specify only the minimum amount of data, use is made of the sign and value of N_2, the type number, for elements which are similar to each other:

(a) $N_2 = +0$ means that the element is identical in all respects to the previous element. In this case only N_1, $+0$, L are specified for the element.

(b) N_2 is written as a positive number for an element whose data is in the normal form given above, the values of the parameters $J, ..., G$ being specified down to the last non-standard value.

(c) N_2 is written as a negative number for an element which has some of its data the same as the previous element. The loading conditions must all be the same if this facility is to be used. It is required to specify only up to and including the last parameter which differs from the previous element, with the exception that A, I and z or b, h_b, d, h_w, f, h_f are considered as groups and if any of the group is different from that of the previous element, then the whole group must be specified. It is permissible to specify only N_1, N_2, L, where N_2 is negative: this example is for an element which is identical in all geometric respects to the previous element but has a different type number. An exception to the general rules is that the A, B and C values for type 11 must be specified immediately after the L, even if they are identical to the previous element.

The linking data and beam data listed above are punched on to one continuous tape, and the calculation can now be carried out.

5.4. Output from computer

When the calculation is completed, the computer produces a tape which, when interpreted, results in a series of tabulated numbers as follows:

(a) hub number, rotation v_x, rotation v_y, deflection w
   (with a separate row for each hub)

(b) hub number, element number, moment M_x, moment M_y, shear force S
   (with a separate row for each end of each element)

Items (a) and (b) are then repeated for each loading condition. For hubs on one or more axes of symmetry or asymmetry, the hub forces and moments due to the image beam are also included so that, for example, moments about an axis of symmetry are zero and the shear forces are doubled.

(c)   For the stresses, a series of six columns is listed, as follows:
   First column: label at end 0 of the beam element
   Second column: z, distance from neutral axis to outermost fibre, positive if measured to bottom flange and negative if measured to top flange
   Third column: σ_0 outer fibre stress at end 0
   Fourth column: σ_B outer fibre stress at end B
   Fifth column: $\sigma_{\text{max.}}$ outer fibre stress at a maximum between 0 and B, if any
   Sixth column: σ_P outer fibre stress at point of application of concentrated load, if any

The stresses in each element are listed for every loading condition before proceeding to the next element. If there are no $\sigma_{\text{max.}}$ or σ_P values, only σ_0 and σ_B are given. If there is a $\sigma_{\text{max.}}$ but no σ_P, only σ_0, σ_B and $\sigma_{\text{max.}}$ are given. If there is a σ_P but no $\sigma_{\text{max.}}$, then σ_0, σ_B, $+0{\cdot}000+0$, σ_P are given.

In the outputs, the majority of the data are given in 'floating point' form, where the number of the index giving the power of ten is printed immediately after each number. An exception is for z, where the decimal point is already inserted ('fixed point' working). The form of the output will be illustrated in a numerical example.

5.5. Numerical example

A 5×2 beam grillage with simply supported edges, shown in fig. 9 (a), will be analysed under the following three separate loading conditions:

(1) uniform loading of 1 lb/in. along all the longitudinal and transverse beams;

(2) concentrated load of 1 lb at the centre of the grillage (point 0, fig. 9 (a));

(3) the combined effect of (1) and (2) (to illustrate the use of the program with a combined loading).

Due to symmetry, only one quarter of the grillage need be considered, and the bottom left-hand quarter has been chosen. The type numbers for the elements are given in fig. 9 (b), and the label numbers for the elements and hubs are shown in fig. 9 (c). The data tape

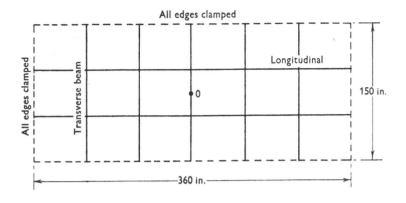

(a)

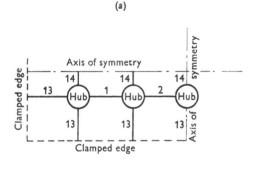

(b)

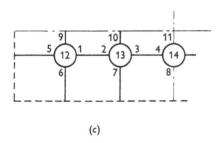

(c)

Fig. 9. Example for solution by displacement method using computer.

(a) Grillage geometry

Beam data: all longitudinals and all transverse beams are evenly spaced.

Longitudinals: $A = 5$ in.^{2} $I = 300$ in.^{4} $z = 2$ in. $J = 0$ ⎱ Torsion
Transverse beams: $A = 2{\cdot}7$ in.^{2} $I = 75$ in.^{4} $z = -3$ in. $J = 0$ ⎰ neglected

(b) Element type numbers

(c) Element and hub label numbers

[53]

is shown in table 11. This data would normally be printed in one continuous column, but, for convenience in presentation, it is shown here as several columns. The following points should be particularly noted:

(i) Elements 7, 8 and 10 are identical in all respects to the previous element in each case, so $N_2 = 0$ has been used.

(ii) Element 3 is identical in all geometrical respects to element 1 but has a different type number.

(iii) Apart from element 11, all elements have the same loading conditions, so that the loading conditions need only be specified for elements 1 and 11. The loading conditions were examined before labelling the elements and it was easily seen to be most convenient to give the last label to the element with the loading different from the rest.

TABLE 11. *Data tape for numerical example in fig. 9*

Linking data		Beam data				
+3	+3	+1	+5	+6	+9	+11
+5	+14	+1	−13	−13	−14	+14
+12	+7	+5	+5	+2·7	+2·7	+2·7
+14	+0	+300	+300	+75	+75	+75
+12	+10	+2	+2	−3	−3	−3
+1	+0	+60	+60	+50	+50	+50
+13	−1	+0	+180	+270	+90	+90
+5	+14	L	L	L	L	L
+0	+4	+1	.	.	.	+1
+6	+13	+0	.	+7	+10	+0
+0	+8	+0	.	+0	+0	+0
+9	+0	+0	.	L	L	+1
+0	+11	+1	.	.	.	+0·5
−1	+0	+0	.	+8	.	+1
+13	−1	.	.	+0	.	+1
+2	+15	+3	.	L	.	+0·5
+12	L	−2	.	.	.	.
		L	.	.	.	.

The results are given in table 12, where it can be seen that some $\sigma_{max.}$ values are produced; these are the only stress values for which the results for loading condition (3) are not the sum of the results for the first two loading conditions.

TABLE 12. *Results for numerical example in fig.* 9

(Edited after printing out, to show labels and symbols.)

LOADING 1		v_x	v_y	w
Hub	+12	+1·08687461 −5	−9·15540164 −6	+5·37116917 −4
Hub	+13	+6·07932748 −6	−1·82813981 −5	+1·07250870 −3
Hub	+14	−3·75542042 −14	−2·12649809 −5	+1·24754555 −3
Hub	+12	M_x	M_y	S
Elements:	+1	+1·52463074 +2	+0·00000000 +0	−3·88649940 +1
	+5	−1·52463135 +2	−3·10608243 −7	+3·94258347 +1
	+6	+7·28808891 −6	+6·15652916 +2	+2·44391730 +1
	+9	+6·31604394 −7	−6·15652817 +2	−2·50000000 +1
Hub	+13			
Elements:	+2	−6·84362518 +2	+0·00000000 +0	−2·11350050 +1
	+3	+6·84362610 +2	+0·00000000 +0	−2·75845470 +1
	+7	+1·70110939 −5	+1·43699257 +3	+7·37195530 +1
	+10	+1·47422500 −6	−1·43699249 +3	−2·50000000 +1
Hub	+14			
Elements:	+4	−2·19395614 −5	+0·00000000 +0	−6·48309059 +1
	+8	+2·01898572 −5	+1·70551498 +3	+8·98308978 +1
	+11	+1·74970488 −6	−1·70551495 +3	−2·50000000 +1
LOADING 2		v_x	v_y	w
Hub	+12	+1·36539386 −8	−6·29197866 −9	+3·69129456 −7
Hub	+13	+2·58400047 −8	−2·75582063 −8	+1·61674813 −6
Hub	+14	−1·32318656 −17	−7·61213541 −8	+2·90714416 −6
Hub	+12	M_x	M_y	S
Elements:	+1	−2·44765043 +0	+0·00000000 +0	−2·06580162 −2
	+5	+2·44765031 +0	+4·98651997 −9	−1·33186509 −2
	+6	+6·70359257 −9	+5·66278268 −1	+3·39766922 −2
	+9	+5·80950361 −10	−5·66278081 −1	+0·00000000 +0
Hub	+13			
Elements:	+2	+1·20816937 +0	+0·00000000 +0	+2·06580176 −2
	+3	−1·20816946 +0	+0·00000000 +0	−1·69472337 −1
	+7	+2·93610236 −8	+2·48023874 +0	+1·48814321 −1
	+10	+2·54450162 −9	−2·48023859 +0	+0·00000000 +0
Hub	+14			
Elements:	+4	−7·73019515 −9	+0·00000000 +0	+3·38944674 −1
	+8	+7·11370340 −9	+6·00921988 −1	+1·61055315 −1
	+11	+6·16491774 −10	−6·00921869 −1	−5·00000000 −1
LOADING 3		v_x	v_y	w
Hub	+12	+1·08824001 −5	−9·16169364 −6	+5·37486045 −4
Hub	+13	+6·10516750 −6	−1·83089563 −5	+1·07412546 −3
Hub	+14	−3·75674361 −14	−2·13411022 −5	+1·25045270 −3
Hub	+12	M_x	M_y	S
Elements:	+1	+1·50015320 +2	+0·00000000 +0	−3·88856525 +1
	+5	−1·50015442 +2	−3·05621711 −7	+3·94125156 +1
	+6	+7·29479234 −6	+6·16219185 +2	+2·44731493 +1
	+9	+6·32185344 −7	−6·16219097 +2	−2·50000000 +1
Hub	+13			
Elements:	+2	−6·83154480 +2	+0·00000000 +0	−2·11143465 +1
	+3	+6·83154541 +2	+0·00000000 +0	−2·77540188 +1
	+7	+1·70404551 −5	+1·43947282 +3	+7·38683681 +1
	+10	+1·47676951 −6	−1·43947273 +3	−2·50000000 +1
Hub	+14			
Elements:	+4	−2·19472917 −5	+0·00000000 +0	−6·44919662 +1
	+8	+2·01969710 −5	+1·70611592 +3	+8·99919534 +1
	+11	+1·75032136 −6	−1·70611586 +3	−2·55000000 +1

TABLE 12 (cont.)

Loading	Element	z	σ_0	σ_B	$\sigma_{max.}$	σ_P
1	+1	+2·000	+1·016 +0	+4·562 +0	+6·051 +0	.
2	+1	+2·000	−1·632 −2	−8·054 −3	.	.
3	+1	+2·000	+1·000 +0	+4·554 +0	+6·040 +0	.
1	+3	+2·000	+4·562 +0	+3·596 +0	+7·099 +0	.
2	+3	+2·000	−8·054 −3	+5·973 −2	.	.
3	+3	+2·000	+4·554 +0	+3·656 +0	+7·122 +0	.
1	+5	+2·000	+1·016 +0	−2·675 +1	.	.
2	+5	+2·000	−1·632 −2	−1·009 −2	.	.
3	+5	+2·000	+1·000 +0	−2·676 +1	.	.
1	+6	−3·000	−2·463 +1	+7·425 +1	.	.
2	+6	−3·000	−2·265 −2	+4·530 −2	.	.
3	+6	−3·000	−2·465 +1	+7·430 +1	.	.
1	+7	−3·000	−5·748 +1	+1·400 +2	.	.
2	+7	−3·000	−9·921 −2	+1·984 −1	.	.
3	+7	−3·000	−5·758 +1	+1·402 +2	.	.
1	+8	−3·000	−6·822 +1	+1·614 +2	.	.
2	+8	−3·000	−2·404 −2	+2·981 −1	.	.
3	+8	−3·000	−6·824 +1	+1·617 +2	.	.
1	+9	−3·000	−2·463 +1	−2·463 +1	−3·713 +1	.
2	+9	−3·000	−2·265 −2	−2·265 −2	.	.
3	+9	−3·000	−2·465 +1	−2·465 +1	−3·715 +1	.
1	+10	−3·000	−5·748 +1	−5·748 +1	−6·998 +1	.
2	+10	−3·000	−9·921 −2	−9·921 −2	.	.
3	+10	−3·000	−5·758 +1	−5·758 +1	−7·008 +1	.
1	+11	−3·000	−6·822 +1	−6·822 +1	−8·072 +1	.
2	+11	−3·000	−2·404 −2	−2·404 −2	+0·000 +0	−5·240 −1
3	+11	−3·000	−6·824 +1	−6·824 +1	+0·000 +0	−8·124 +1

5.6. Discussion

The program described for the Pegasus computer is capable of analysing plane grillages of great generality, which makes it necessary to solve for three unknowns at each intersection except on axes of symmetry where one or more of the slopes is known. This limits the size of grillage which can be analysed to about twenty intersections if there is no symmetry, and correspondingly more with symmetry. The same type of program, worked out for a larger capacity computer, could tackle more intersections. Both shear and torsion terms are included. If it is desired to neglect shear deflections, either G or A can be taken very large, say 10^6 times their actual values: for $N_1 > 0$, a large A is the best choice, but for $N_1 < 0$, a large G is the only choice since the program then takes $A = dh_w$. However, it must be remembered that a large G will produce a large GJ unless J is reduced proportionately or is zero. To neglect the torsional stiffness, J should be taken as zero.

It should be remembered by those not familiar with digital computers, that a program developed for one computer will not, in general, work on another computer, though the logic of the method and specification of the input and output data can be utilized. The program developed for Pegasus has been described in some detail, to illustrate what is involved in this type of numerical solution. Some detailed points relating to this program, such as the possibility of hubs not lying at an intersection point, and the operating instructions for the machine, are not discussed here. Likewise, the specialist work of programming, which was carried out to produce the program tape, is not described.

All the straightforward grillage solutions which reduce to solving a large set of simultaneous equations are eminently suitable for programing for a computer. In orthogonal grillages where torsion is neglected, the slopes can be eliminated for each beam, which reduces the number of unknowns to one per intersection point and increases to three times the number of intersections which can be tackled. An example of a treatment involving only one unknown per intersection is the Force Method described in chapter 2, and it would be most useful to write a program based on this method.

Chapter 6

EFFECTS OF SHEAR AND TORSION

In chapters 2–4, it was assumed that the effects of shear deflections and torsional rigidity of the beams are sufficiently small to be neglected. These assumptions were made for simplicity. To allow for shear introduces additional terms in the expressions for deflection, and slope discontinuities at the intersections. Likewise, the presence of torsional rigidity introduces further terms in the moment equilibrium equations. Many of the methods described could not be used at all if shear and torsion were included. The most suitable method for examining shear and torsion is the Slope Deflection Method, solving the equations on a digital computer, as described in chapter 5.

6.1. Effect of shear deflections

It is well known that, in structures of beams, shear deflections only become significant for deep beams having comparatively short spans, and it has been suggested that the shear term might be more important in grillages than in other beam structures, since there are comparatively short spans between the intersections. Exact numerical analysis of two grillages, now to be described, has shown that this contention is unfounded. In these calculations torsion was neglected.

The first grillage, with three longitudinal and nine transverse beams, has a plan form similar to that in fig. 4, and measures 50 ft × 16 ft 6 in. The longitudinal girders are 6 in. × 12 in. I beams (BSB 122 giving $I = 300$ in.^{4}, $A_{\text{web}} = 5$ in.^{2}) and the transverse stiffeners are 4 in. × 9 in. I beams (BSB 115 giving $I = 75$ in.^{4}, $A_{\text{web}} = 2 \cdot 7$ in.^{2}). The structure is simply supported and is subjected to a central concentrated load of 10 tons. For calculating the shear deflection terms, it was assumed that the shear force is uniformly resisted by the webs of the members and that the flanges do not carry shear. The calculated deflection and bending moments at the central intersection point are given in table 13. It may be seen that, although the total deflection was increased by $7\frac{1}{2}$% including shear, the effect on the bending moments was negligible.

TABLE 13. *Bending moments: maximum deflection in*
3 × 9 beam grillage under central concentrated load

Quantity	Shear included	Shear neglected
Deflection, in.	0·155	0·144
Girder bending moment, tons in.	153	153
Stiffener bending moment, tons in.	100	101

The second grillage, shown in fig. 10, is an idealization of a caisson at the entrance to a lock and is subjected to hydrostatic pressure. The validity of the idealization might be considered too dubious for accurate results, but the structure provides a valuable illustration of the effect of shear in grillages composed of exceptionally deep beams. Again, the shear area was taken as the area of web. The calculated maximum deflection and bending moments are given in table 14. In this case, the total deflection was increased by 43 % due to shear, yet the bending moments were only affected by up to 3 %.

The results from these two calculations show that, although the deflections may be significantly increased by including the shear deflections, the bending moments are only marginally affected, and so the effect on the stresses may be neglected for design purposes. This may be explained in general terms by the fact that the shear deflection is usually very roughly proportional to the bending deflection for all the intersection points in a grillage, and so does not significantly affect the equations of compatibility and hence the bending moments.

TABLE 14. *Bending moments: maximum deflection in*
caisson under hydrostatic pressure

Quantity	Shear included	Shear neglected
Deflection, point 4, in.	0·426	0·299
Horizontal bending moment, point 4, tons in.	$98·4 \times 10^3$	100×10^3
Vertical bending moment, point 1, tons in.	$15·1 \times 10^3$	$15·6 \times 10^3$

6.2. Effect of torsional rigidity

It was considered likely that torsional rigidity would be more important in grillages having a wide difference in stiffness between the longitudinal and transverse members. Then, the torsional moments

[59]

on the stiffer members might become comparable with the bending moments on the less stiff members, and the torsion terms would be significant in the equations of equilibrium. The type of beam section also has an important bearing on the torsional inertia J. For I beams and similar open cross-sections, J is equal to $\Sigma\, bt^3/3$, where b = width and t = thickness of the web or flange, and the summation includes the web and both flanges. For thin-walled tubes with closed cross-sections, J is equal to $4A^2/\int ds/t$, where A is the area enclosed by the walls, and the integral is taken round the perimeter s of the cross-section. Numerically, the values of J are appreciably higher for closed

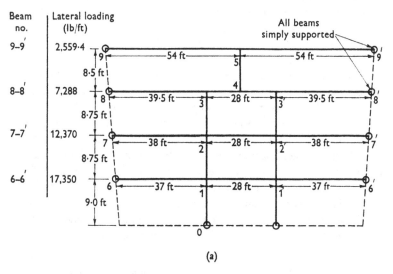

(a)

Fig. 10. Idealization of caisson structure. (a) Elevation;

cross-sections. Two grillage examples to examine torsion will now be described, and in each case shear is also included taking the shear area equal to the area of the web.

The first grillage chosen to investigate the effect of torsional rigidity is a 'uniform' grillage containing four equal and evenly spaced longitudinals and five equal and evenly spaced transverse beams. The structure measures 20 ft × 8 ft 4 in., and is simply supported at all four edges. The following two alternative beam structures are considered:

(a) *Beams of open cross-section*

Longitudinals 10 in. deep × 5 in. wide I beams (overall) with 0·72 in.

[60]

Beam	Section	Moment of inertia (in.² ft.)	Shear area (in.²)
9–9′	9 in × $\frac{7}{28}$ in. · $\frac{17}{40}$ in. · 25 in. × 1 in. · 13·8 in.	19,630	140·6
8–8′	21 in. × $\frac{17}{40}$ in. · $\frac{17}{40}$ in. · 50 in. × 1 in. · 9·8 ft · 13·8 ft	29,690	140·6
7–7′	21 in. × $\frac{17}{40}$ in. · 5 ft × $\frac{1}{2}$ in. · 50 in. × 1 in. · $\frac{17}{40}$ in. · 13·8 ft	29,503	151·3
6–6′	10·5 in. × $\frac{17}{40}$ in. · 50 in. × 1 in. · 6 ft × $\frac{17}{40}$ in. · 5 ft × $\frac{17}{40}$ in. · 10 ft 8 in.	25,672	86·7
0–3′	21 in. × $\frac{17}{40}$ in. · 25 in. × 1 in. · $\frac{17}{40}$ in. · 13·7 ft	19,400	139·5
4–5′	25 in. × 1 in. · $\frac{15}{40}$ in. · 13·7 ft	17,000	123·0

(b)

Fig. 10. Idealization of caisson structure. (b) beam sections.

thick flanges and 0·36 in. thick web, $I = 174·1$ in.⁴, $J = 1·38$ in.⁴. Transverse beams 2·75 in. deep × 1·75 in. wide I beams (overall) with 0·375 in. thick flanges and 0·2 in. thick web. $I = 2·00$ in.⁴, $J = 0·067$ in.⁴.

(b) *Beams having a rectangular closed box section*

Longitudinals 10 in. deep × 5 in. wide (overall) with 0·72 in. thick flanges and 0·36 in. thick webs. $I = 193·0$ in.⁴, $J = 115·1$ in.⁴. Transverse beams 2·75 in. deep × 1·75 in. wide (overall) with 0·375 in. thick flanges and 0·2 in. thick webs. $I = 2·13$ in.⁴, $J = 1·69$ in.⁴.

The following two loadings are separately considered:

(i) Uniform pressure of 5 lb/in.². For the purposes of the grillage calculation, this is equivalent to a concentrated load of 4000 lb at each intersection point. There is a small additional local bending moment term which can be calculated by the methods described in chapter 11.

(ii) Concentrated load of 5 tons applied at the centre of the grillage.

This position lies on a beam, but is midway between intersection points. Thus, four grillage calculations are given and these are carried out first including, then ignoring, torsional rigidity.

TABLE 15. *Maximum deflection and stresses in 4 × 5 beam grillage under uniform pressure*

Beam sections ...	I beams		Box beams	
Torsion	Included	Neglected	Included	Neglected
Central deflection, in.	0·800	0·803	0·663	0·722
Maximum longitudinal stress, tons/in.²	8·88	8·93	7·03	8·12
Maximum transverse stress, tons/in.²				
Grillage term	13·38	13·30	15·42	11·96
Local bending term	− 0·37	− 0·37	− 0·44	− 0·34
Total	13·01	12·93	14·98	11·62

The results for the uniform pressure loading are shown in table 15. It can be seen that, for the grillage constructed using I beams, torsion has affected the results to a negligible extent, but for the grillage with box beams the effect of including torsion has been to reduce the deflection by 8 % but to increase the maximum stress occurring in the light transverse beams by 29 %. The maximum stress in the longitudinals is reduced by 13 %. The results for the concentrated load are shown in table 16. Again, torsion has a negligible effect in the I-beam structure, but is quite significant for the box beams. In the latter structure, the effect of torsion has reduced the deflection by 11 % and the maximum stress in the light transverse

TABLE 16. *Maximum deflection and stresses in* 4×5 *beam grillage under central concentrated load*

Beam Sections ...	I beams		Box beams	
Torsion	Included	Neglected	Included	Neglected
Central deflection, in.	0·213	0·214	0·171	0·192
Longitudinal stress, tons/in.²	3·18	3·20	2·44	2·91
Transverse stress, tons/in.²	17·67	17·95	12·79	16·57

beams by 23 %. The largest stress in the longitudinals is reduced by 15 %.

The second grillage chosen to investigate torsional rigidity is a 'uniform' grillage with four equal and evenly spaced longitudinals and four equal and evenly spaced transverse beams. In this case, both the longitudinal and transverse beams are identical to the longitudinals in the 4×5 beam grillage in the last example, and calculations are carried out for I beams and for box beams. The grillage measures 12 ft 6 in. square and is acted on by a uniform pressure of 20 lb/in.² which is equivalent to a concentrated load of 18,000 lb at each intersection point. In this example, where the two sets of beams are of equal stiffness, the local bending term can be ignored.

TABLE 17. *Maximum deflection and stress in symmetrical* 4×4 *beam grillage under uniform pressure*

Beam sections ...	I beams		Box beams	
Torsion	Included	Neglected	Included	Neglected
Central deflection, in.	0·431	0·431	0·366	0·379
Maximum stress, tons/in.²	11·86	11·87	10·37	10·72

The results are shown in table 17, and, as in the previous example, torsion has a negligible effect for I beams. For the box-beam structure, the effect of including torsion is to reduce both the maximum deflection and stress by about 3·4 %. Thus, the effect is appreciably less than for the first grillage which has widely differing longitudinal and transverse members.

The results from both these examples show that torsion can be

[63]

neglected in unplated grillages containing beams with open cross-sections such as I beams, channel sections and T sections. When these sections are attached to the continuous plating of plated grillages, the torsional inertia J is somewhat greater than the value from $\Sigma bt^3/3$, due to lateral bending of the flanges caused by the centre of rotation being transferred to a position near the point of attachment to the plating. However, even in these plated grillages, torsion is of no practical significance. The only case where torsion should be included is in grillages where the members have closed box or tubular sections, and the effect of torsion is greatest if the longitudinal and transverse members have widely differing stiffnesses. Then, the effect on the stresses in the lighter members can be quite marked, their maximum stress being either lowered or raised by torsion, according to the configuration and the loading; the effect on the maximum stress in the heavier members and on the deflection is to reduce these values, but to a less marked degree.

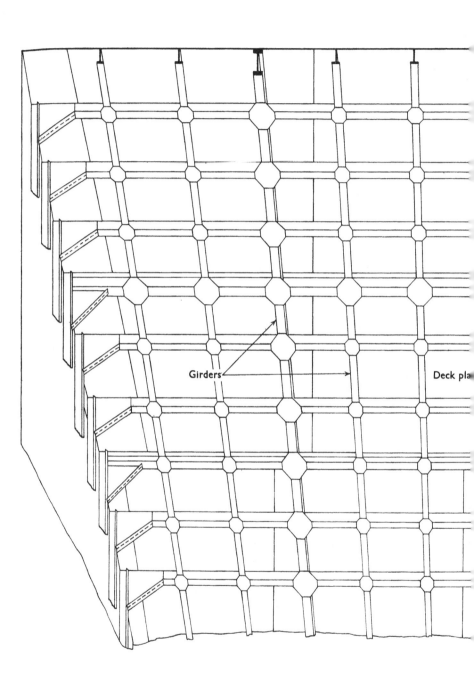

Girders ←   →   Deck pla

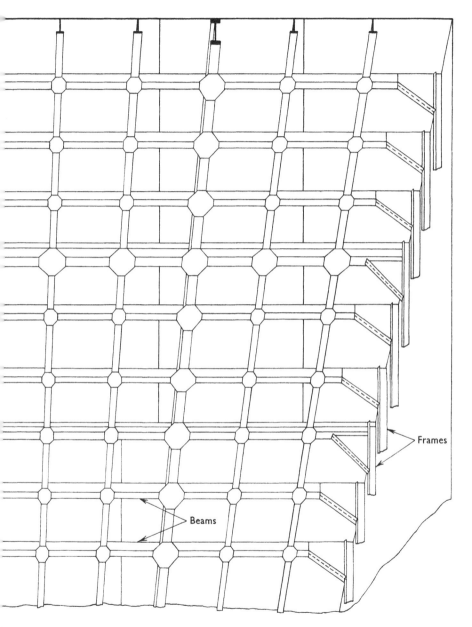

Frames

Beams

Fig. 11. Flight deck structure of aircraft carrier showing stiffening on underside.

Chapter 7

SHIP GRILLAGES

Historically, grillages have been introduced in ship structures in order to stiffen the plating which forms the decks, sides, bottom and bulkheads of the structure. With the increasing size of vessels, successively more stiffening has proved necessary to provide adequate strength and stiffness, so that, nowadays, the stiffening members are often thought of as the major part of the structure. Nevertheless, the weight of the plating usually exceeds that of the stiffening, and must be taken into account in investigations aimed at producing a minimum weight design.

The view looking inside a compartment of the main hull of a frigate, shown in fig. 1, illustrates both flat and curved grillages. The T-section stiffeners are attached to the plating, by welding, which forms a virtually perfect shear connection. Fig. 11 shows the flight deck of an aircraft carrier, designed to withstand the concentrated loadings due to landing of an aircraft at any random position.

The primary function of the plating in present-day ship grillages is to provide watertight integrity, and it is designed to resist the lateral pressure forces which are applied normal to its surface. The loading may then be regarded as transferred to the beam structure. The plating also has a most important secondary function, since it acts as a flange to the beams. This chapter is concerned with the way in which the loading is transferred from the plating to the beam structure, and the contribution of the plating to the strength and stiffness of the beams.

In principle, the problem of a plated grillage may be solved using the equations of linear elasticity. The deflections and stresses in each plate panel are governed by the well-known partial differential equations:

$$\nabla^4 \phi = 0, \tag{7.1}$$

$$\nabla^4 w = q/D, \tag{7.2}$$

where ϕ is the Airy stress function defining the stresses in the plane of the plate, w is the bending deflection, q is the lateral pressure, $D = Eh^3/12(1-\mu^2)$ and h = thickness. The beams may be analysed by the usual Euler–Bernoulli theory, and the boundary conditions for the solutions to (7.1) and (7.2) are obtained by specifying equilibrium and compatibility at each beam position. Unfortunately, in any practical grillage, the large number of plate panels renders this type of treatment virtually impossible numerically, even with aid of a modern digital computer.

An attempt at a simpler theoretical solution has been made by Kendrick, who considered concentrated loadings applied directly over the beams. For this case, it was possible to ignore the flexural rigidity of the plating, and to consider only the stresses in the plane of the plate given by equation (7.2). It was further assumed that the 'in plane' displacements u and v could be obtained from the solution of simpler equations of the form

$$u_{xx} + \frac{(1-\mu)}{2} u_{yy} = 0. \tag{7.3}$$

The justification was that almost identical stresses and deflections were obtained from equations (7.1) and (7.3) for structures having single direction stiffening. When applied to the orthogonal grillage, equation (7.3) and its counterpart in v lead to independent plating stresses and displacements in the two orthogonal directions, apart from the effect of the vertical reactions at the intersections. The numerical results from this analysis were compared with answers from an unplated grillage calculation including the plating as a beam flange of effective breadth equal to the beam spacing, acting with both the longitudinal and transverse beams. The two methods gave almost identical values for the maximum deflections and stresses. This suggests that a simple approach of taking the full beam spacing as effective breadth may be appropriate, though Kendrick's analysis was by no means rigorous.

Since mathematical analysis of plated grillages is fraught with difficulty, an experimental approach has been adopted, to obtain design rules for the effective breadth of plating to be incorporated into unplated grillage calculations, and for the effective loading on the beams. The measurements have in some instances included a detailed survey of the stress distribution in the plate panels, leading to results as in fig. 12: the effective breadth of plating may then be obtained from the integral of the heart of the plate stress over half a beam spacing each side of the beam, divided by the value opposite the beam. The moment of resistance of a beam with this effective flange is identical to that of the actual composite beam-plating structure. Unfortunately the strain-gauge measurements, leading to fig. 12, are very time-consuming and costly, and also the mid-surface

stresses obtained are not particularly accurate due to the large
and rapidly varying bending component of stress. The results for
effective breadth have shown a fair amount of scatter, even within a
single grillage. Fortuitously, this is of little practical significance,
since the deflections and stresses in the true beam flanges remote
from the plating, which are usually the largest stresses, are quite
insensitive to changes in effective breadth. This is because, for

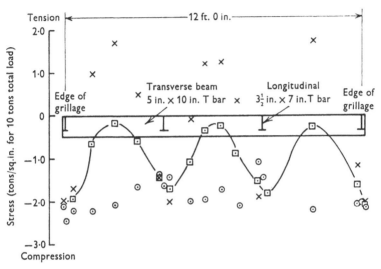

Fig. 12. Grillage no. 1 loaded at $P1$ (see fig. 14). Plating stresses at right
angles to transverse cross-section 8 in. from centre line of grillage. $\odot$,
stress on top surface of plating; $\times$, stress on lower surface of plating; $\boxdot$, heart
of plate stress—average.

practical scantlings, the neutral axis lies close to the plating, and
changes in the breadth of plate flange have only a small effect on the
moment of inertia and section modulus, particularly the latter, as
in fig. 13. The approach adopted to determine the effective breadth
has been to test a number of grillages having a similar structure and
loading, and measure the deflections and beam flange stresses.
Grillage calculations taking a number of effective breadths of plating,
assumed constant for the longitudinals and for the transverse beams,
have shown what breadth gives the best overall agreement. Likewise,
the effective loading on the beam structure, where significant, has
been determined largely by trial and error, to obtain the best agreement between the calculated and measured stresses.

5-2

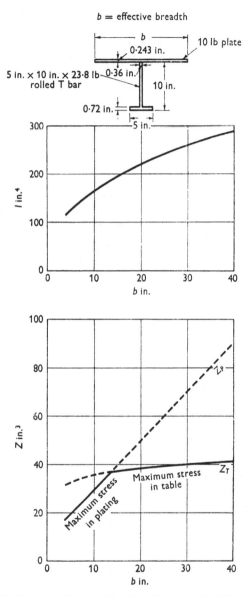

Fig. 13. Moment of inertia I and section modulus Z for T bar with plate flange.

7.1. Concentrated loadings

Three welded structures in mild steel have been tested, the details and load positions being shown in figs. 14 and 15. Load positions over the beams were chosen, since the largest stresses in the flanges are

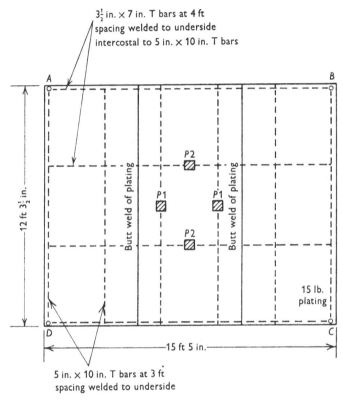

Fig. 14. Scantlings and load positions for grillage nos. 1 and 3. (The two grillages have identical scantlings.) *ABCD*, support positions. Two equal loads at positions *P*1—grillage nos. 1 and 3. Two equal loads at positions *P*2—grillage no. 1 only.

caused by loading directly over these members. Historically, grillages nos. 1 and 2 were first tested, and the effective breadth of plating was found to be about half the beam spacings. Detailed examination of the measurements, which included a survey of the stresses in the plate panels, indicated that the reduced effective breadth was due to the initial dishing of the plate panels due to welding which was of the order of $\frac{1}{4}$ in. A third grillage was therefore constructed, similar to the first, and the plate panels were made flatter by hammering the welds at the toe of the T bars with a caulking tool. In this way, the dishing was reduced to within $\frac{1}{20}$ in. This last grillage gave deflections which

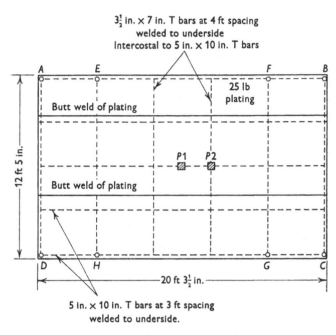

Fig. 15. Scantlings and load positions for grillage no. 2. Supports at $EFGH$, load at position $P1$. Supports at $ABCD$, load at position $P2$.

were generally about 10 % smaller than for grillage no. 1, and an effective breadth intermediate between 50 and 100 %, as shown in fig. 16. The maximum stresses in grillage nos. 1 and 2 are compared with the calculated values for an effective breadth 100 and 50 % of the beam spacing in table 18. The agreement is generally good and illustrates how insensitive the calculated values are to changes in the effective breadth. It may be concluded that, for calculations of singly plated grillages of the type tested under concentrated loadings, the effective breadth should be taken as half the beam spacing.

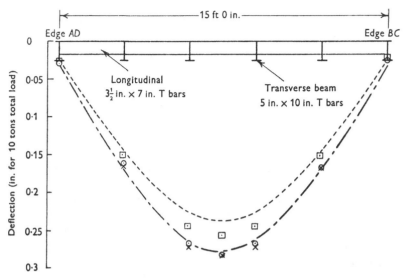

Fig. 16. Grillage no. 1, loaded at $P1$ (see fig. 14). Deflections of inner longitudinal beams: results for grillage no. 3 also shown. - - - -, Calculation taking effective breadth of plating = beam spacing; — · —, calculation taking effective breadth of plating = 50 % beam spacing; ×, experimental points (beam on side AB of centre line); ⊙, experimental points (beam on side CD of centre line); ⊡, experimental points for grillage no. 3.

TABLE 18. *Maximum stresses, tons/in.^{2} for 10 tons total load*

Grillage number	Support positions (figs. 14 and 15)	Load positions (figs. 14 and 15)	Longitudinal (L) or Transverse (T) beam stress	Calculated stress Effective breadth 100 % beam spacing	Effective breadth 50 % beam spacing	Measured stress
1	$ABCD$	$P1$	L	5·91	6·00	6·4, 6·6
			T	2·31	2·29	2·6, 2·68
		$P2$	L	8·15	8·24	9·2, 9·3
			T	1·22	1·26	1·26, 1·38
2	$EFGH$	$P1$	L	4·30	4·23	4·68, 4·74
			T	6·01	5·97	5·3, 5·7
	$ABCD$	$P1$	L	5·66	5·60	5·38
			T	4·55	4·49	4·38, 4·90
		$P2$	L	4·06	4·02	4·30
			T	5·46	5·41	5·90, 6·26

[71]

7.2. Concentrated loadings on structures stiffened in one direction only

In grillages where the beams of one set are much more closely spaced, and have a smaller stiffness than the beams of the intersecting set, the latter beams along the short panel edges may often be assumed to be rigid against deflection. Then the structure of the less stiff beams may be analysed separately from the overall grillage behaviour. For a single concentrated load applied at any random position, each beam should be capable of withstanding the load applied midway between intersections, and it may be considered as a beam continuous across the heavy transverse beams, which have effectively zero deflection. The effective breadth of plating may be taken as half the beam spacing. The actual maximum deflection and stress at the load position will be less than the values obtained from this calculation, due to transfer of load to the adjacent unloaded beams by bending and 'in-plane' stresses in the plating. The amount of load transferred depends on the relative stiffness of plating to beams, but might amount to as much as 40 % in heavily plated structures.

A theoretical analysis based on the equations of linear elasticity, equations (7.1) and (7.2), has been carried out by Clarkson for the case of simply supported ends, shown in fig. 17. This analysis proved rather tedious numerically, and it has been followed up by experimental work, first at a small scale in xylonite, and more recently at near full scale in welded mild steel. The latter test structures were of 100 in. span, with at least six plate panels ranging from 20 in. to 40 in. wide. The range of scantlings included rolled T-section stiffeners from $1\frac{3}{4}$ in. $\times$ $4\frac{1}{2}$ in. to 4 in. $\times$ 8 in. and plating from 0·19 in. to 0·47 in. thick. The experimental results for the maximum deflection and stress agreed quite well with the earlier theoretical answers, and the following empirical formulae were devised to fit all the available data:

$$W = \frac{\text{Max. deflection of structure}}{\text{Max. deflection of single beam (including effective breadth half-beam spacing)}}$$
$$= 1\cdot44 - 0\cdot34 \left(\frac{a}{b}\right)^{\frac{1}{2}} - 2\frac{bD}{EI}; \tag{7.4}$$

$$S = \frac{\text{Max. stress of structure}}{\text{Max. stress of single beam (including effective breadth half-beam spacing)}}$$
$$= 0\cdot88 - 0\cdot025 \left(\frac{a}{b}\right)^{3}\frac{bD}{EI}. \tag{7.5}$$

It should be noted that equations (7.4) and (7.5) are given in terms of I, the moment of inertia of a beam without an effective breadth of plating. The parameter bh/A, giving the amount of carry over of load by the in-plane stresses in the plating, does not appear explicitly, since for a given class of structures such as the ones tested there is a very rough one-to-one relationship between bh/A and bD/EI (the

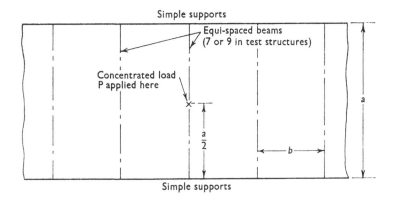

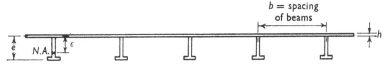

Fig. 17. Structure and notation.

Notation:

A = cross-sectional area.
I = moment of inertia about its own neutral axis. } For beam section alone without plating.

E = Young's modulus.
μ = Poisson's ratio. } Beams and plating assumed to be of same material.

bending stiffness parameter). The formulae were based on experiments within the ranges $a/b = 2.5\text{--}5$ and $bD/EI = 0.001\text{--}0.07$, and they should not be applied for other proportions without further theoretical or experimental investigation. It is considered that the formulae may be used to calculate the carry-over effect for beams continuous over the ends instead of simply supported, taking the span a as the length between points of contraflexure, though this has yet to be checked experimentally.

7.3. Uniform pressure

Three welded structures in mild steel have been tested, and in each case the grillage consisted of heavy transverse beams intersected by very much lighter longitudinal stiffeners. Calculations had indicated that this type of arrangement would lead to a minimum weight grillage with the largest stress in the longitudinals and transverse beams roughly equal, and this rather surprising stress distribution was confirmed by the measurements. Owing to the comparatively light section of the longitudinals, their stresses were critically dependent on the local sagging action between intersections, and consequently on the effective loading transferred to the beam structure by the plating. For each test structure design, two identical grillages were placed back to back and connected by thin low alloy steel plate, to obtain virtually simply supported edges. This edge connection was designed so that its stretch under the reactions from the grillage beams was negligible compared with the deflection of the grillage, yet its bending stiffness was only a few percent of that of the grillage. In the test assemblies, one grillage was placed with its stiffeners facing outside, and the other inside. The complete box was welded together to be pressure tight and loaded inside by water pressure. A typical test assembly is shown in plate I. The structures measured $162\frac{1}{2}$ in. $\times$ 130 in. and 200 in. $\times$ 100 in., and the main stiffeners were either 5 in. $\times$ 10 in. or 6 in. $\times$ 12 in. T bars. Full-scale construction techniques were used throughout, and the maximum working load pressure was approximately 20 lb/in.^{2}.

For comparison with the measurements, grillage calculations were carried out for effective breadths equal to 30 plate thicknesses, 50 thicknesses and half the beam spacings. Shear deflection terms were also included and amounted to about 20 % in one case. There was a marked difference in the measured deflections for grillages with the plating forming mainly the compression flange to the stiffeners (stiffeners facing outwards in the test assembly) and grillages with the plating as the tension flange (stiffeners facing inwards), particularly in the more thinly plated structures. The effective breadths giving best agreement were 30 plate thicknesses for plating in compression and half the beam spacings for plating in tension. The low effective breadths for plating in compression were associated with a moderately large permanent set of the panels after loading, up to $\frac{1}{2}$ in. deep, whereas, for tension, the dishing was merely that initially built in due to welding, usually appreciably less than $\frac{1}{2}$ in.

PLATE I

Uniform pressure test assembly no. 3 showing deformation of top grillage
after test to $41 \cdot 5$ lb/in.^{2}.

(*Facing p* .74)

The stresses of the heavy transverse beams exhibited a smooth distribution (fig. 18), but the stresses of the longitudinals showed marked local bending (fig. 19). It was not possible to tie up these local bending moments theoretically on the basis of a uniform line load per unit length. Instead, a load distribution in the form of a series of parabolic waves, varying from zero at each intersection to a maximum midway between, was assumed, and the amplitude

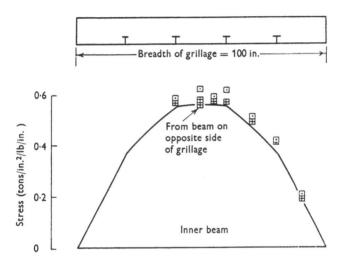

Fig. 18. Uniform pressure grillage no. 3: stresses on tables of inner transverse beams. *Rest of legend as fig.* 19 (page 76).

chosen to fit the experimental stresses. It was found that 0·75 or 0·85 times the total pressure loading should be considered as applied to the longitudinals (longer panel edges) for aspect ratios of 1·25 or 2 respectively. The remainder of the total loading should be taken as uniformly applied along the transverse beams. Grillage calculations using these assumptions gave excellent agreement with the measurements, as indicated in figs. 18 and 19.

It should be appreciated that only a limited amount of experimental data is available, and that the rules for effective breadth and loading to be used in design calculations should be applied with caution, outside the range of structures tested.

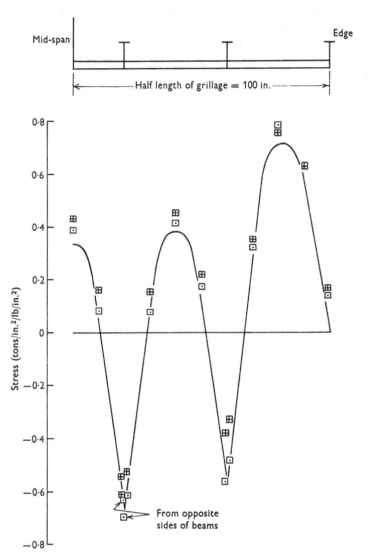

Fig. 19. Uniform pressure grillage no. 3: stresses on tables of inner longitudinal beams. ⊡, Experimental points, top grillage loaded on plating side; ⊞, experimental points, bottom grillage loaded on stiffener side; ———, theoretical curve for effective breadth = half-beam spacings.

7.4. Thick plating

All the experimental work described has been concerned with plating less than $\frac{3}{4}$ in. thick, and plate panels ranging from 20 in. to 48 in. wide, and the effective breadth is taken as half the beam spacing or 30 plate thicknesses, according to the type of loading and stress distribution. These rather small effective breadths are associated with small permanent sets in the plate panels due to either welding deformation or the loading. In grillages containing plating 1 in. or more in thickness, particularly where the beam spacings are not proportionately increased, the welding deformation may be negligible, and generally larger effective breadths are appropriate.

Chapter 8

WELDED CONNECTIONS

In the theoretical discussions, it has been assumed that reactions in the form of shear forces are transferred between the webs of intersecting members without any relative moment occurring. Likewise, it is usually assumed that the strength of the grillage beams is unimpaired by local failures at the intersections. These assumptions are only approximately true, and it is the task of the designer to ensure that the joints have adequate stiffness and strength for the particular loadings envisaged. A deficiency in the stiffness of joint could cause increased bending moments in the adjoining members compared with those calculated, and a reduction in strength of the larger member due to cut-out holes would have obvious adverse effects on the overall strength. Soft joints might also give rise to increased plate panel deflections rendering the plating less effective as a flange to the stiffeners. Although failures at the joints, due to inadequate detail design, are seldom catastrophic (except in cases of suddenly applied dynamic loadings), they might give rise to extremely inconvenient and costly repairs. It should be realized that the failures could occur due to either the single application of a very high loading, or the repeated application of smaller loads causing fatigue.

In the days of riveted structures, it was necessary to place quite heavy angle connections between the intersecting members. Nowadays, simpler welded connections are used, though a large diversity of arrangements has been utilized from time to time. To assist the designer in his choice of connections, a series of tests has recently been carried out on connections between a main 4 in. × 8 in. T-bar beam with a smaller T-bar intersecting member. A wide range of designs was tested, with the smaller member passing through a hole cut in the web of the main member, and the details of the specimens for a 3 in. × 6 in. piercing member are given in fig. 20. A reference beam was made with an accurately prepared butt-welded two-piece insert collar fitted around the piercing T bar. Great care was taken throughout to select material of uniform yield stress. The specimens were tested statically well beyond the elastic limit, with a 'cruciform' arrangement, a downward load at the ends of the main member being reacted upwards at the ends of the piercing member as shown

[78]

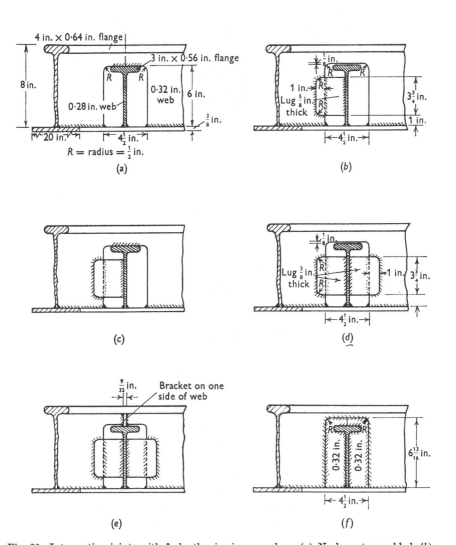

Fig. 20. Intersection joints with $\frac{3}{4}$ depth piercing member. (a) No lugs, top welded; (b) one lug; (c) one lug and top welded; (d) two lugs; (e) two lugs and bracket (or top welded); (f) reference beam. Two-piece collar plate.

in fig. 21. Deflexions were measured at the mid-span of each member, and the results are shown in figs. 22 and 23. To compare the strengths of the main members, the loads corresponding to a departure of $\frac{1}{2}$ in. from perfect elastic behaviour were read from the graphs and expressed as percentages of the reference beam value, giving an

[79]

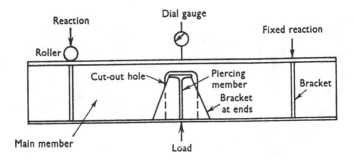

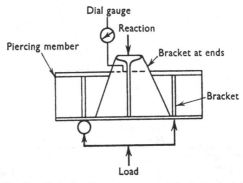

Fig. 21. Cruciform loading arrangement for tests on joints.

'efficiency' η_1. Similarly, to provide a comparison between the rigidities of the connections, the loads to cause $\frac{1}{10}$ in. relative deflection were expressed as percentages of the reference beam value, giving a second efficiency, η_2 (extrapolation was required for joints A, C and E). The results are summarized in table 19, and it can be seen that there are significant differences between the arrangements studied. Lug connections without a bracket or a direct weld between the piercing member flange and main member web give a poor joint rigidity, whereas the arrangements with no lugs or one lug, which have an open web hole up to the full depth of the slot, give a reduced strength of main member. A combination of two lugs, and bracket or welded flange to web connection, may be regarded as very satisfactory, and should be adopted for all major structures which are

[80]

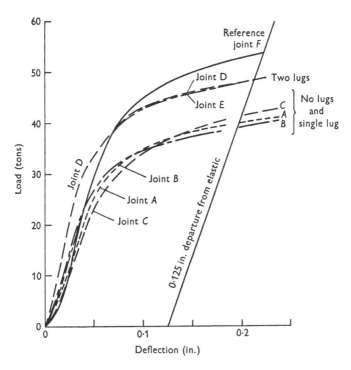

Fig. 22. Deflection at mid-span of main member with
$\frac{3}{4}$ depth piercing member.

TABLE 19. *Structural efficiency of welded connections*

Efficiency	Main member, η_1		Joint, η_2	
$\dfrac{\text{Depth passing member}}{\text{depth main member}}$	0·75	0·56	0·75	0·56
Joint A: no lug plus bracket or direct weld	75	90	75	85
Joint B: one lug	75	75	35	50
Joint C: one lug plus bracket or direct weld	75	90	80	90
Joint D: two lugs	90	90	50	80
Joint E: two lugs plus bracket or direct weld	90	—	95	—
Joint F: two-piece insert collar —reference beam	100	100	100	100

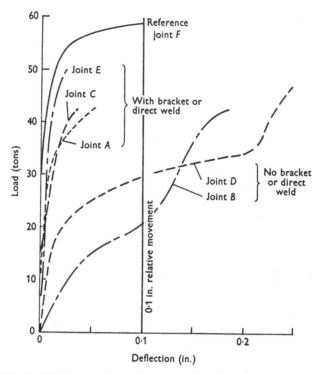

Fig. 23. Relative movement at joint with $\frac{3}{4}$ depth piercing member.

strength critical. Suitable sizes for the lugs and brackets are given as proportions of the area of web of the piercing member in table 20. (It should be noted that the proportions are smaller for the smaller depths of piercing member, since the lengths of the cut-out hole in the main member, the unsupported plating and the lugs are all reduced, and the effect of the hole is appreciably smaller.) The alternative connection consisting of a two-piece lapped collar, used as a reference for comparing the test results, would not normally be adopted in practice, due to the expensive fitting involved. Although the experiments have only been carried out for symmetrical T-section stiffeners, the order of merit given by the 'efficiencies' η_1 and η_2 should be applicable for other sections.

The case of connections between stiffeners of equal depth has not, so far, been investigated experimentally. If rolled sections are being used, one set of stiffeners has to be cut and made intercostal, and it is suggested that the most satisfactory connection would be to make

TABLE 20. *Cross-sectional areas of lugs and brackets for connections with two lugs plus bracket or direct weld*

(A_w = area of web of piercing member. For diagram showing details of connection see fig. 20(e).)

	Depth of piercing member/depth of main member		
	$\geqslant 0\cdot65$	$0\cdot65$ to $0\cdot40$	$\leqslant 0\cdot40$
Each lug (vertical area)	$0\cdot5A_w$	$0\cdot4A_w$	$0\cdot3A_w$
Bracket (in way of weld to main member web)	$0\cdot8A_w$	$0\cdot65A_w$	$0\cdot5A_w$

direct weld connections between the two flanges and the two webs, and cover the flange of the intercostal member with a rider plate spanning the intersection. In the case of fabricated stiffeners, only the webs of one set have to be made intercostal. The flanges of both the longitudinal and transverse members should be continuous, with one set passing over the other and fillet welds round all four edges of the overlapping rectangle. In special circumstances, where the small difference in depth in this arrangement cannot be tolerated, the welds connecting the intercostal flanges to the intersections should be full penetration welds with a double- or single-edge preparation.

The work described has not included a study of repeated loadings and the effect of stress concentrations at the connections on their fatigue strength. Quite high-stress concentrations may exist in the main members at the edge of the cut-out hole, which might lead to early fatigue failures. Until quantitative data are available, the choice of connection will have to be based on the static test results, although where very high repeated loadings are anticipated, the two-piece lapped collar should be adopted.

Chapter 9

DATA SHEETS FOR SIMPLY SUPPORTED GRILLAGES UNDER A SINGLE CONCENTRATED LOAD

Even with assistance from modern electronic computers, the analysis of successive trial designs may prove a tedious process, and, for this reason, design data sheets giving the maximum bending moments have been prepared for several ranges of grillage. The case of a single concentrated load, which might be applied at any position on the grillage, is applicable to the design of floors to withstand heavy weights, bridges under wheel loads, the decks of ships to withstand items of heavy machinery, and to the flight decks of aircraft carriers for the wheel loads during aircraft landings.

The number of possible grillage configurations is very large, and it hardly seems feasible to present data for grillages having non-equal beams or non-even spacings. The data presently available is for 'uniform' grillages, in which all the longitudinals are equal and evenly spaced, likewise all the transverse beams. In addition, only grillages containing an odd number of beams in each direction are considered and it is assumed that the boundaries are simply supported (fig. 24).

In general, the largest stresses in a simply supported grillage are not obtained for loading at the centre. For the grillages in fig. 24, larger stresses are usually obtained for load positions 1 and 2, near the centre but midway between beam intersections. Even these positions may not give the largest stresses, but in cases so far investigated other positions have given either lower or only slightly larger stresses. For the present work, the most severe loading positions have been assumed to be at points 1 and 2. An exception has been made for grillages with only one beam in either set (p or $q = 1$): as for the larger numbers of beams, it was assumed that loading the central beam of either set is the most severe, but the position of loading across the span of the central transverse beam, which is supported only at the central longitudinal and the ends, has been calculated to give the largest bending moment. This position lies between about one-fifth and half-way across the span, depending on the relative stiffnesses of the longitudinal and transverse beams.

[84]

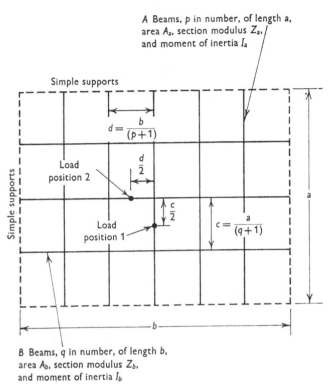

Fig. 24. Uniform grillage showing nomenclature and loading positions. Bending moment at position 1 due to concentrated load P, at position $1 = \alpha Pc$. Bending moment at position 2 due to concentrated load P, at position $2 = \beta Pd$. Stiffness ratio, $\lambda = b^3 I_a / a^3 I_b$. A grillage containing p 'A' beams and q 'B' beams is referred to as a $p \times q$ grillage. Thus a 7×5 grillage contains 7 'A' beams and 5 'B' beams.

The maximum bending moment data are given in figs. 25–33 for grillages containing 1, 3, 5, 7, 9 or 15 beams in each direction, and for values of the stiffness ratio $\lambda = b^3 I_a / a^3 I_b$ between 1 and 100, with p (the number of A beams of length a and moment of inertia I_a) $\geqslant q$ (the numer of B beams of length b and moment of inertia I_b). This means, in effect, that to lie within the range of the data sheets, the larger number of beams should have the greater stiffness. With this data it is possible to study the minimum weight design of grillages having the larger number of beams spanning the shorter width, which is a very common type of configuration. The data were obtained before digital electronic computers were generally available, by the Fourier Series Method using hand computing.

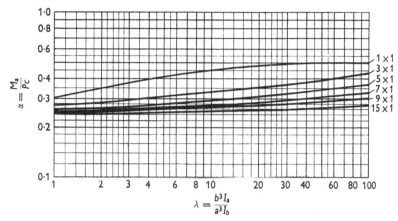

Fig. 25. $p \times 1$ grillages—maximum bending moment in A beams.

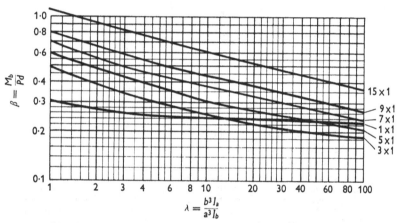

Fig. 26. $p \times 1$ grillages—maximum bending moment in B beams.

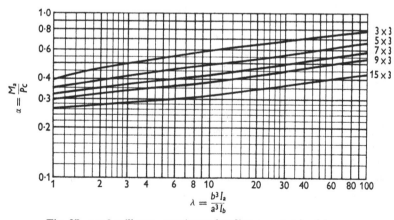

Fig. 27. $p \times 3$ grillages—maximum bending moment in A beams.

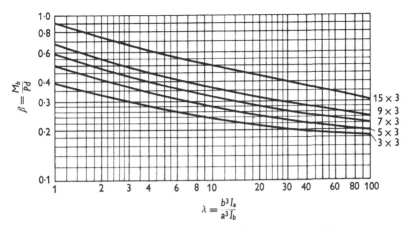

Fig. 28. $p \times 3$ grillages—maximum bending moment in B beams.

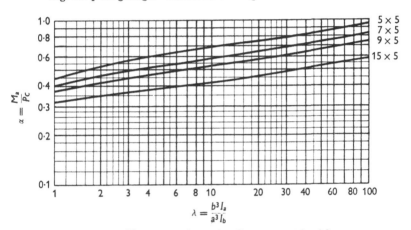

Fig. 29. $p \times 5$ grillages—maximum bending moment in A beams.

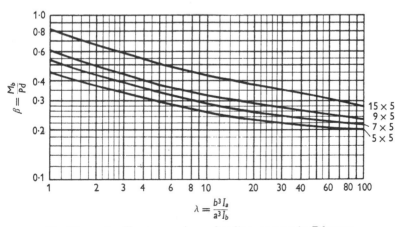

Fig. 30. $p \times 5$ grillages—maximum bending moment in B beams.

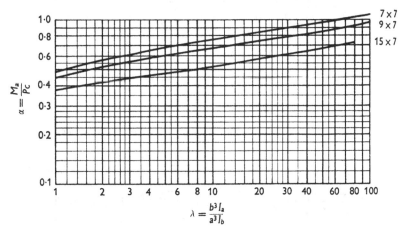

Fig. 31. $p \times 7$ grillages—maximum bending moment in A beams.

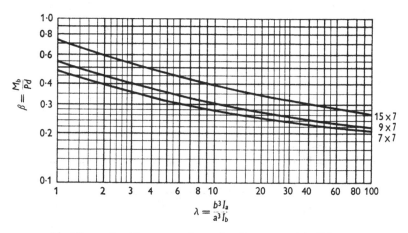

Fig. 32. $p \times 7$ grillages—maximum bending moment in B beams.

[88]

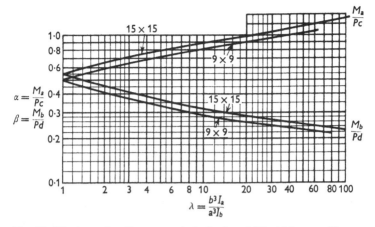

Fig. 33. Maximum bending moments in 9×9 and 15×15 beam grillages.

Chapter 10

MINIMUM WEIGHT DESIGN FOR CONCENTRATED LOADS

Using the bending moment design data for simply supported grillages under a concentrated load given in chapter 9, the proportioning of material between the longitudinal and transverse members of a grillage to obtain a minimum weight design may now be considered. In the discussion which follows, it is assumed that the number of stiffening members in each direction has been chosen *a priori*. Possibly this has been determined by considerations other than transverse strength, as in ships' decks and bottom structures which have to withstand fairly large compressive stresses associated with longitudinal bending of the hull. If the lateral loads were the only important factor, successive calculations could be carried out for various stiffener spacings, to determine the minimum weight design.

The design of the plate panels, to determine the plate thickness for a given panel size, is a question which does not need to be specifically considered to determine the moment of inertia ratio of the stiffeners, I_a/I_b, for minimum weight. Nevertheless, in many designs, the plating provides a large proportion of the total weight of structure, and this will be illustrated in a numerical example. The choice of a suitable panel size may significantly affect the total weight of structure.

10.1. Relation between area, moment of inertia and section modulus

In order to carry out weight calculations, relations between the area, moment of inertia and section modulus of the beams are required.

For rectangular beams measuring jd wide by d deep, the following relations hold:

$$\left.\begin{array}{ll} A = jd^2, & A = 6^{\frac{2}{3}}j^{\frac{1}{3}}Z^{\frac{2}{3}}, \\ I = \tfrac{1}{12}jd^4, \quad \text{i.e.} & I = \dfrac{6^{\frac{1}{3}}}{2j^{\frac{1}{3}}}Z^{\frac{4}{3}}. \\ Z = \tfrac{1}{6}jd^3, & \end{array}\right\} \tag{10.1}$$

In general, the beams used in grillages do not have geometrically similar cross-sections. However, for certain classes of beams, the condition of geometrical similarity leading to relations of the form

$$A = hZ^m, \quad I = kZ^n \tag{10.2}$$

is satisfied to a reasonable degree of accuracy. Thus, for British Standard Joists, the relations

$$A = 1{\cdot}480Z^{0{\cdot}549}, \quad I = 1{\cdot}007Z^{1{\cdot}458} \tag{10.3}$$

provide a reasonable approximation for the entire range of joists available in B.S. Specification No. 4, 1932, as is shown in figs. 34 and 35. For beams consisting of Admiralty welding T bars with an associated plate flange, similar relations hold approximately between the area of T bar alone, A, and the moment of inertia I and section modulus Z of T bar with associated plate flange, and the constants in equations (10.2) for two thicknesses and a range of widths of plating are given in table 21. The agreement for the various sizes of T bar with these formulae is shown for a plate flange 24 in. wide and 0·367 in. thick (15 lb) in figs. 36 and 37.

TABLE 21. *Relations between area of cross-section of beam alone, and section modulus and moment of inertia of beam with associated plate flange for various 'families' of beams*

$(A = hZ^m, I = kZ^n.$ 40 lb plate has been taken as 0·98 in. thick.)

Beams	Plating flange		h	m	k	n
	Wt./ft² (lb)	Width (in.)				
Square cross-sections	—	—	3·30	0·667	0·909	1·333
Rectangular cross-sections b wide × $2b$ deep	—	—	2·62	0·667	1·145	1·333
British Standard Joists no plating flange	—	—	1·480	0·549	1·007	1·458
Admiralty welding T bars	30	60	0·652	0·624	2·39	1·369
		37·5	0·656	0·625	2·52	1·336
		24	0·662	0·624	2·55	1·313
		22·5	0·670	0·624	2·56	1·312
	15	60	0·71	0·606	2·40	1·339
		24	0·700	0·618	2·48	1·279
		18·75	0·700	0·626	2·56	1·252
		11·25	0·703	0·628	2·43	1·234

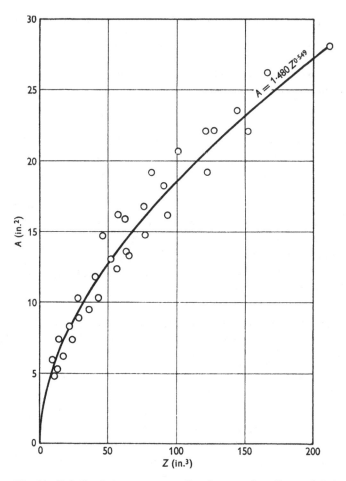

Fig. 34. Relation between cross-sectional area and section modulus
for British Standard Joists.

10.2. Formulae for weight of grillage

It is first necessary to ascertain whether the largest stress occurs in
the A or B beam structure. If this stress is in an A beam, we have the
condition

$$\frac{\alpha Pc}{Z_a} < \frac{\beta Pd}{Zb},$$

using the notation given in fig. 24. This equation may be rewritten in
terms of the moments of inertia, using relations of the form of equations

[92]

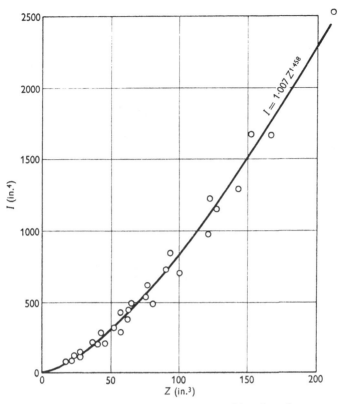

Fig. 35. Relation between moment of inertia and
section modulus for British Standard Joints.

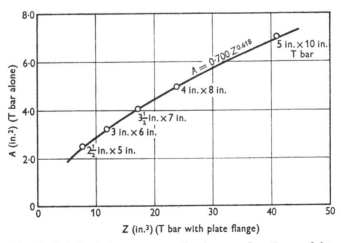

Fig. 36. Relation between cross-sectional area and section modulus
for Admiralty Welding T bars with 15 lb plate flange 24 in. wide.

[93]

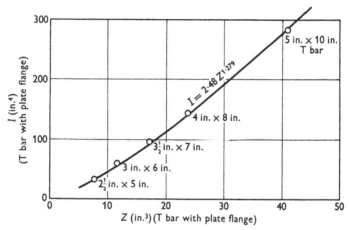

Fig. 37. Relation between moment of inertia and section modulus for Admiralty Welding T bars with 15 lb plate flange 24 in. wide.

(10.2) with suffixes a and b to denote the A and B beams respectively, giving

$$\frac{(p+1)a}{(q+1)b}\left(\frac{k_a I_b}{k_b I_a}\right)^{1/n_b} \frac{\alpha}{\beta}\left(\frac{k_a}{I_a}\right)^{(1/n_a)-(1/n_b)} > 1. \qquad (10.4)$$

The modulus of section Z_a required to limit the maximum A beam stress to a value σ_s under the design load P is

$$Z_a = \alpha P c / \sigma_s.$$

The area and moment of inertia of the A beams follow simply, using equations (10.2),

$$A_a = h_a \alpha^{m_a}\left(\frac{Pc}{\sigma_s}\right)^{m_a}, \quad I_a = k_a \alpha^{n_a}\left(\frac{Pc}{\sigma_s}\right)^{n_a};$$

i.e.

$$I_b = \frac{I_b}{I_a} k_a \alpha^{n_a}\left(\frac{Pc}{\sigma_s}\right)^{n_a},$$

$$A_b = h_b \left(\frac{k_a I_b}{k_b I_a}\right)^{m_b/n_b} \alpha^{m_b n_a/n_b}\left(\frac{Pc}{\sigma_s}\right)^{m_b n_a/n_b}.$$

The total volume of material $V = p a A_a + q b A_b$, and substituting the appropriate areas we obtain

$$W = \frac{V}{p a h_a \left(\dfrac{Pa}{4\sigma_s}\right)^{m_a}}$$

$$= \left(\frac{4}{q+1}\right)^{m_a}\left\{\alpha^{m_a} + \frac{qb}{pa}\frac{h_b}{h_a}\left(\frac{k_a I_b}{k_b I_a}\right)^{m_b/n_b} \alpha^{m_b n_a/n_b}\left(\frac{Pc}{\sigma_s}\right)^{(m_b n_a/n_b)-m_a}\right\},$$

$$(10.5)$$

where W is a non-dimensional weight parameter.

[94]

If, on the other hand, the largest stress occurs in a B beam, the weight of structure, derived by a similar argument, is

$$W = \left(\frac{4}{q+1}\right)^{m_a} \left\{ \left(\frac{(q+1)\,b}{(p+1)\,a}\right)^{m_a\,n_b/\,n_a} \left(\frac{k_b I_a}{k_a I_b}\right)^{m_a/n_a} \beta^{\,m_a\,n_b/n_a} \left(\frac{Pc}{\sigma_s}\right)^{(m_a\,n_b/\,n_a)-m_a} \right.$$
$$\left. + \frac{qb}{pa}\left(\frac{(q+1)\,b}{(p+1)\,a}\right)^{m_b} \frac{h_b}{h_a} \beta^{\,m_b} \left(\frac{Pc}{\sigma_s}\right)^{m_b-m_a} \right\}.$$
$$(10.6)$$

For grillages in which the A and B beams belong to the same family of sections, as in unplated grillages of British Standard Joists, or in grillages of T bars with square panels of plating between the beams, $h_a = h_b = h, m_a = m_b = m$, etc., and equations (10.4) to (10.6) are appreciably simplified, giving

$$\frac{(p+1)\,a}{(q+1)\,b}\left(\frac{I_b}{I_a}\right)^{1/n}\frac{\alpha}{\beta} > 1,\qquad(10.7)$$

$$W = \left(\frac{4\alpha}{q+1}\right)^{m}\left\{1+\frac{qb}{pa}\left(\frac{I_b}{I_a}\right)^{m/n}\right\}\qquad(10.8)$$

for grillages with the maximum stress in the A beam structure and

$$W = \left(\frac{4\beta}{q+1}\right)^{m}\left\{\frac{(q+1)\,b}{(p+1)\,a}\right\}^{m}\left\{\left(\frac{I_a}{I_b}\right)^{m/n}+\frac{qb}{pa}\right\}\qquad(10.9)$$

for grillages with the maximum stress in the B beam structure. The formulae (10.4) to (10.9) enable the weight to be quickly calculated, once values have been assigned for a, b, p, q and I_a/I_b.

10.3. Equal stress condition

It may be seen from equations (10.4) to (10.9) that the number of variables determining the weight of grillage is very large, particularly when the two sets of beams (including plating flanges) belong to different families of sections, as in grillages of T bars with other than square panels of plating between the beams. In this case, the weight depends on a parameter (Pc/σ_s), containing the magnitude of loading and size of grillage: this is due to the differing dimensions of h_a and h_b, k_a and k_b when the beams belong to different families.

To examine how the magnitude of Pc/σ_s and the differing A and B beams may affect the design for minimum weight, a fairly wide range of calculations has been made for 3×3, 15×3 and 15×15 beam grillages having ratios of length to breadth $b/a = 1\cdot0$, $1\cdot5$ and $3\cdot0$.

In these calculations, the associated plate flange was included in the moments of inertia and section moduli but not in the cross-sectional areas and weights, since for each size of grillage and beam configuration the weight of plating has a constant value. For each case, the weight of grillage stiffening required to limit the largest stress to the design value was calculated for a range of values of I_a/I_b. Some typical graphs of results are shown in figs. 38–40. As

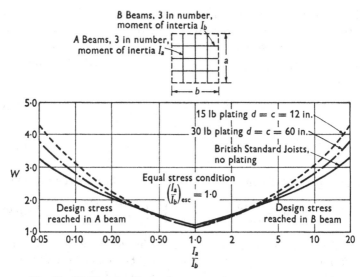

Fig. 38. Weight of grillage, showing minimum at the equal stress condition: (i) 3×3 grillages with $b/a = 1 \cdot 0$.

might be expected, the minimum weight design was obtained from a structure which has the same stress per unit load in the A beam for a load at position 1, as in the B beam for a load at position 2 (fig. 24). This is a well-balanced design, the structure being subsequently referred to as the 'equal stress condition' structure, and the weight parameter and moment of inertia ratio for this condition are denoted W_{esc} and $(I_a/I_b)_{esc}$. It may be seen from the graphs that the moment of inertia ratio $(I_a/I_b)_{esc}$ varies somewhat according to the type of beams used and the relative values of h, k, m and n, but W_{esc} is little affected.

Since there are only a limited number of beam sizes commercially available, and there may be fairly wide intervals between two adjacent sizes, it is usually necessary in practice to depart slightly from the equal stress condition. The curves in fig. 39 show that, for equal numbers of beams in each direction, it would be less expensive in weight to err towards stiffer transverse (shorter) beams, giving a design with the maximum stress in the longitudinals. If there is a larger number of transverse beams than longitudinals, the design

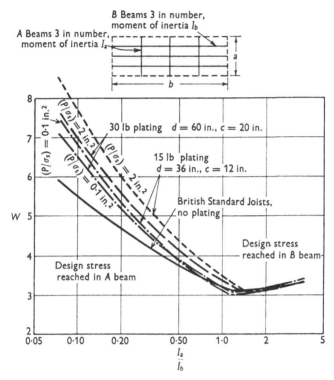

Fig. 39. Weight of grillage, showing minimum at the equal stress condition: (ii) 3×3 grillages, with $b/a = 3 \cdot 0$.

should err towards stiffer longitudinals, giving a design with the maximum stress in the transverse beams. A general guide, based on the curves in figs. 39 and 40, and data for other grillage configurations (not given here) is that the design should err towards extra stiffness of the beams which are spaced more widely apart. It should be recognized that the parameter W does not include the plating weight, and the effect of an unbalanced design of beam structure on the total weight is much less marked than is indicated by W.

Since the values of W_{esc} are only very slightly affected by the type of beams used and the values of h, k, m and n, it was decided to determine the equal stress condition for British Standard Joists only. The results for $(I_a/I_b)_{esc}$ for a wide variety of grillage configurations are

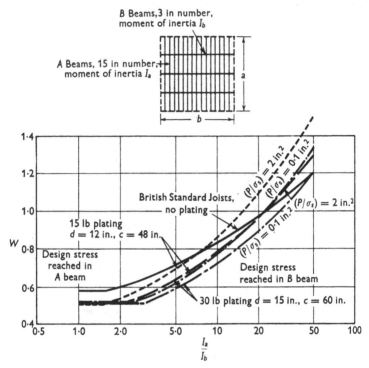

Fig. 40. Weight of grillage, showing minimum at the equal stress condition: (iii) 15×3 grillages, with $b/a = 1.0$.

given in table 22, and the corresponding values of W_{esc} are shown in fig. 41. Noting that fairly small changes in beam dimensions give rise to significant changes in the moment of inertia, it will be seen that, for all the cases studied, the longitudinal and transverse beams are of very similar size, to obtain a well-balanced design satisfying the equal stress condition.

TABLE 22. *Design for equal stress condition: moment of inertia ratio, $(I_a/I_b)_{esc}$ for grillages of British Standard Joists*

Grillage $p \times q$	Grillage, b/a				
	1·0	1·25	1·5	2·0	3·0
1 × 1	1·0	1·11	1·03	0·85	0·55
3 × 1	1·56	1·84	1·83	1·74	1·44
5 × 1	2·28	2·31	2·23	2·01	1·74
7 × 1	2·90	2·70	2·47	2·19	1·90
9 × 1	3·22	3·03	2·74	2·38	2·07
15 × 1	4·02	3·53	3·23	2·87	2·43
3 × 3	1·0	1·48	1·62	1·61	1·26
5 × 3	1·19	1·74	1·93	1·81	1·60
7 × 3	1·30	1·78	1·87	1·88	1·76
9 × 3	1·49	1·78	1·84	1·90	1·84
15 × 3	1·66	1·73	1·72	1·79	1·75
5 × 5	1·0	1·65	1·93	1·80	1·50
7 × 5	1·13	1·74	1·93	1·93	1·67
9 × 5	1·12	1·71	1·82	1·88	1·76
15 × 5	1·23	1·55	1·69	1·74	1·89
7 × 7	1·0	1·75	1·96	1·89	1·67
9 × 7	1·07	1·77	1·98	1·95	1·79
15 × 7	1·10	1·57	1·72	1·86	1·92
9 × 9	1·0	1·79	2·06	2·04	1·77
15 × 15	1·0	1·82	2·13	2·20	1·96

10.4. Comparison of weights of grillage and structure with only transverse beams

It might be expected that for large ratios of length to breadth of structure b/a, a lighter structure would be obtained by incorporating only the transverse (shorter) beams. This will be so when the transfer of load by the longitudinal beams from the loaded transverse beam to the adjacent beams is not sufficient to compensate for the weight of the longitudinals.

The structure with only transverse beams must be designed so that each individual beam can withstand the load P applied at its mid-span. There is some transfer of load to adjacent beams by plate bending and by forces in the plane of the plating, as described in in chapter 7, but for the time being this correction term is neglected. Thus, the modulus of section required for each transverse (A) beam assuming simple supports, is

$$Z_a = Pa/4\sigma_s.$$

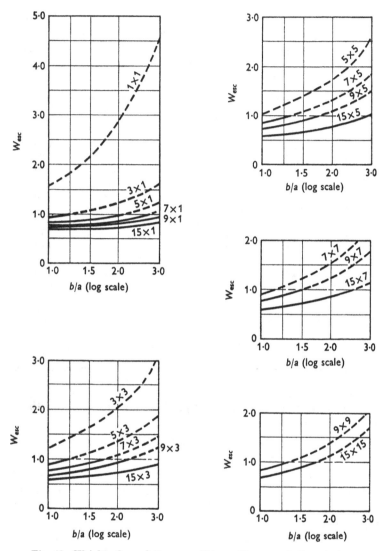

Fig. 41. Weight of equal stress condition grillages excluding plating.

From equations (10.2), the corresponding area is $A_a = h_a(Pa/4\sigma_s)^{m_a}$ and the total volume of material is given by

$$V_s = paA_a = pah_a(Pa/4\sigma_s)^{m_a},$$

or

$$W_s = \frac{V_s}{pah_a(Pa/4\sigma_s)^{m_a}} = 1. \tag{10.10}$$

It may be seen that the weight parameter W, first introduced in equation (10.5), is equal to the weight of grillage stiffening divided by the weight of a single set of beams, p in number, across the shorter span, designed to carry the same load. The physical significance of the curves in fig. 41 may now be seen. The portions shown as full lines, for $W_{esc} < 1$, indicate the range where the grillage gives a lighter structure than single direction stiffening. Likewise, the dotted portions, for $W_{esc} > 1$ indicate the range where single direction stiffening is lighter. The values of W ignore the weight of plating which must be greater for single direction stiffening, and they also ignore the effect of transfer of load to adjacent beams by the plating in single direction stiffened structures which would make this arrangement relatively lighter. Thus, the exact point at which single direction stiffening becomes lighter than the grillage is still a little nebulous, and both arrangements should receive detailed attention if W_{esc} lies in the range between about 0·75 and 2.

10.5. Design procedure

(1) Choose the number of longitudinal and transverse beams. This determines the thickness of plating required for either a grillage with orthogonal stiffening or the corresponding structure with single direction stiffening. The design of the plate panels is a subject in itself, and is not specifically considered in this text. References are given in the Bibliography.

(2) Read W_{esc} from fig. 41. For W_{esc} less than 0·75, only the grillage should be considered further. For $0·75 < W_{esc} < 2$, both the grillage and single direction stiffening should be considered. For $W_{esc} > 2$, only single direction stiffening need be considered.

(3) To design the grillage, read $(I_a/I_b)_{esc}$ from table 22. Then calculate $\lambda = b^3 I_a/a^3 I_b$ and obtain M_a and M_b from the bending-moment data sheets in chapter 9. Knowing the maximum design stress σ_s, the required section moduli follow, and stiffener scantlings may be chosen. The total weight of structure is

$$w = \rho(paA_a + qbA_b) + w_p,$$

where w_p is the weight of plating.

(4) To design the single direction stiffened structure, first calculate $Z_a = Pa/4\sigma_s$ which enables a first choice of scantlings to be made. The actual size of stiffener required will be somewhat smaller, but the stress carry-over factor S may be calculated from these first chosen

[101]

scantlings using equation (7.5). (Note that the notation for equation (7.5), given in fig. 17, differs from the grillage notation in the present chapter.) Then, the section modulus required is $S(Pa/4\sigma_s)$ and new scantlings may be chosen. In extreme cases of very heavily plated structures loading to a low value of S, it would be necessary to repeat the calculation of S using the new stiffener scantlings. Once a design of stiffener has been obtained which is compatible with the applied bending moment after allowing for its carry-over factor, the weight of structure may be calculated from

$$w = \rho p a A_a + w_p.$$

Chapter 11

DATA SHEETS FOR UNIFORM PRESSURE

Design data sheets, similar to those presented in chapter 9 for concentrated loads, have also been prepared for uniform pressure loading. The notation is given in figs. 24 and 43, with the following additional notation:

K_a = edge clamping coefficient for A beams
K_b = edge clamping coefficient for B beams

$\left\{\begin{array}{l} K = 0 \text{ for simple} \\ \quad \text{supports,} \\ K = 1 \text{ for clamped} \\ \quad \text{ends,} \end{array}\right.$

$\bar{p}$ = uniform pressure,
$\bar{q}_a$ = local load per unit length on A beams, maximum value midway between intersections,
$\bar{q}_b$ = local load per unit length on B beams, maximum value midway between intersections.

As previously, only 'uniform' grillages containing odd numbers of beams are specifically considered, but a wider range of parameters has been covered, as follows:

$$p = 3, 5, 7, 9 \text{ and } 15, \qquad K_r = 0 \text{ and } 1,$$
$$q = 1, 3, 5, 7 \text{ and } 9, \qquad K_s = 0 \text{ and } 1.$$
$$\lambda = 1/100 \text{ to } 100,$$

To evaluate the data sheets, a total of 832 separate grillage calculations was carried out, using the Force Method, programed for the Pegasus digital computer.

11.1. The two components of grillage bending moments

Uniform pressure loading applied to a plated grillage may be regarded as transferred by the plating to the longitudinal and transverse stiffening members in differing proportions, according to the aspect ratio c/d of the plate panels. In chapter 7, some experimental work was briefly described, from which it was concluded that the effective loading on the beams is in the form of a series of parabolic waves, varying from zero at each intersection to a maximum midway between. It was found that 0·75 or 0·85 times the total pressure loading

[103]

should be considered as applied to the longer panel edges for aspect ratios of 1·25 or 2 respectively. It may be argued from symmetry that half the total loading is applied to both the longitudinal and transverse panel edges for square panels, and an empirical curve fitting these data is given in fig. 42.

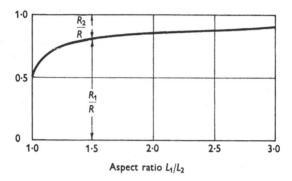

Fig. 42. Loading on beams in grillages under uniform pressure. L_1 = length of plate panel; L_2 breadth of plate panel; $R = \bar{p}L_1L_2$ = total load on plate panel; R_1 = proportion of R applied to long edges; R_2 = proportion of R applied to short edges; $\bar{q}_1 = 1 \cdot 5 R_1/L_1$ = load per unit length on long edges: maximum value midway between intersections; $\bar{q}_2 = 1 \cdot 5 R_2/L_2$ = load per unit length on short edges: maximum value midway between intersections; $\bar{q}_1$ and $\bar{q}_2$ are denoted $\bar{q}_a$ or $\bar{q}_b$ when applied to the A or B beams of a grillage.

Local bending of the grillage beams between their intersections under the loadings described may be quite significant for the lighter beams in grillages with widely differing longitudinal and transverse beams, and this introduces an additional parameter c/d into the total grillage bending moments. Fortunately, the grillage data sheets can be presented without including the local bending moments, and these data will continue to apply when further information becomes available concerning the effective local loadings. The arguments leading to the form of loading used for the data sheets will now be described.

The grillage is first considered to be rigidly supported against deflection at all its intersection points, and the individual beams are analysed as continuous beams under the appropriate local loadings. Bending-moment diagrams for this local bending action for a clamped ended beam, and for a simply supported beam with two, four and six spans, are shown in fig. 43. For design calculations, the following bending-moment values are usually sufficiently accurate:

[104]

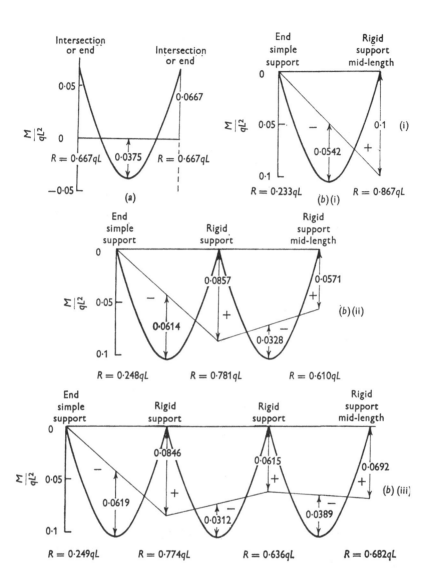

Fig. 43. Bending-moment diagrams for continuous beams on non-deflecting supports under parabolic load on each span. (a) Span of clamped ended beam. (b) Beam with simply supported ends: (i) two spans; (ii) four spans; (iii) six spans. L = distance between intersections or supports; q = load per unit length— maximum value midway between intersections; M = bending moment; R = reaction at support.

Clamped ended beam, at an intersection: $+0.067\bar{q}L^2$ (exact to two figures).

Beam with simply supported ends:

    (a) midway between end and first intersection: $-0.062\bar{q}L^2$;

    (b) at intersection adjacent to end: $+0.085\bar{q}L^2$;

    (c) at any other intersection: $+0.067\bar{q}L^2$.

In the step just described, constraint reactions were applied to prevent the intersections from deflecting. For beams with clamped ends, these reactions are equal to $\bar{p}cd$ at every intersection, due to the symmetry of local bending over each span, whereas for simple supports the reactions are not quite equal, particularly towards the ends (fig. 43). However, a large number of calculations has shown that good accuracy, quite adequate for design calculations, may be obtained taking equal loads $\bar{p}cd$ at every intersection, even for simple supports. Therefore, to complete the solution, the grillage is allowed to deflect under a load $\bar{p}cd$ at each intersection. This is the grillage calculation resulting in the uniform pressure data sheets, now to be presented. These data sheets do not contain the parameter c/d, and the local bending-moment term, described earlier, has to be added to the results obtained.

There is one exception where it is not sufficiently accurate to assume a concentrated load $\bar{p}cd$ at every intersection. This is the case where the grillage contains beams of only two spans ($q = 1$). Then, the intersection load arising from the local loading on the A beams is $0.867\bar{q}_a c$ (fig. 43) and the total load is $0.867q_a c + 0.667\bar{q}_b d$, the values of $\bar{q}_a$ and $\bar{q}_b$ being obtained from fig. 42. Therefore, the data sheet deflection and bending moments should be multiplied by the factor

$$(0.867\bar{q}_a c + 0.667\bar{q}_b d)/\bar{p}cd.$$

11.2. Position of maximum bending moment

For any set of grillage beams which is simply supported, regardless of the boundary condition for the orthogonal set, the maximum bending moment often occurs at the mid-length of a beam, and, where the two sets of beams have similar stiffnesses, the maximum lies in the central beam, that is, at the centre of the grillage. As the number of beams in each set and their stiffnesses become more unequal, the position of the largest bending moment in the stiffer beams moves away from the central beam but still lies at the mid-length.

The position of the largest total bending moment including the local bending term in the less stiff set of beams in grillages with a

[106]

fairly wide difference in stiffness is not easy to specify when these beams are simply supported. The largest value might well arise at any intersection or even at approximately midway between intersections, depending on the magnitude of the local bending term. A careful survey of the bending moments printed out from the computer calculations for the data sheets, in conjunction with the local bending moments from fig. 43 has shown that the following data sheet bending moments need to be considered: in all cases these are negative moments:

(a) Largest bending moment at an intersection adjacent to the edge of the grillage (denoted by $i = 1$). This bending moment is reduced by the local bending term: $+0 \cdot 085\bar{q}L^2$ to be added.

(b) Largest bending moment at a position midway between the edge of the grillage and the first intersection (denoted by $i = \frac{1}{2}$). This is equal to half the value at $i = 1$ mentioned under (a). It is increased by the local bending term: $-0 \cdot 062\bar{q}L^2$ to be added. This is not exactly the largest value in the span, but it is sufficiently close to the maximum for design calculations.

(c) Largest bending moment at an intersection not adjacent to the edge of the grillage (denoted by $i \neq 1$). This is reduced by the local bending-moment term: $+0 \cdot 067\bar{q}L^2$ to be added.

In a few cases, the grillage bending moment at an intersection not adjacent to the edges may become very small negative or even slightly positive, and the total bending moment, not included in cases (a) to (c) above, is approximately equal to the local bending term $+0 \cdot 067\bar{q}L^2$. However, in all such cases examined, either (a) or (b) has proved to be larger.

For a set of grillage beams clamped at its ends, the largest total bending moment always occurs at the clamped edges, usually but not necessarily at the ends of the central beam. As the number of beams in each set and their stiffnesses become more unequal, the position of the maximum bending moment in the stiffer set of beams moves away from the central beam. It should be noted that the position of the maximum total bending moment is not affected by the local bending term since this is equal at all intersections.

11.3. Presentation of data sheets

As described in chapter 3, for grillages containing a large number of beams, the deflections and bending moments may be obtained to a fair accuracy by the orthotropic plate approximation for a large number of beams in both directions, or by the distributed reactions approximation for a large number of beams in one direction only. Where these approximations are valid, the deflections and bending moments no longer depend on the discrete number of beams in the grillage.

For the data sheets being discussed, it was decided that accuracy to within 5% is good enough for design purposes. Therefore, the general method of presentation is to plot the results calculated for 15 or 9 beam grillages in terms of the orthotropic plate or distributed reactions parameters:

(a) Orthotropic plate:

Independent variable (stiffness ratio) $\quad \mu = \dfrac{(p+1)\,b^3 I_a}{(q+1)\,a^3 I_b}$.

Dependent variables $\quad w\bigg/\sqrt{\dfrac{(q+1)\,\bar{p}cda^3}{EI_a}}$

$$M_a/(q+1)\,\bar{p}cda \quad M_b/(p+1)\,\bar{p}cdb.$$

(b) Distributed reactions (for large number of A beams):

Independent variable (stiffness ratio) $\quad \epsilon = \dfrac{(p+1)\,b^3 I_a}{a^3 I_b}$.

Dependent variables $\quad w\bigg/\sqrt{\dfrac{\bar{p}cda^3}{EI_a}}$

$$M_a/\bar{p}cda \quad M_b/(p+1)\,\bar{p}cdb.$$

Where the results for a small number of beams differ by more than about 5% from the approximate orthotropic plate or distributed reactions plot (derived from the 15 or 9 beam results but considered to be accurate for an infinitely large number of beams) these are shown as separate curves on the graphs.

Throughout this work, the results are plotted assuming that the A beams are the more numerous set ($p \geqslant q$) and the detailed presentation is as follows:

(a) Central deflection. Plotted in terms of the orthotropic plate parameters (figs. 44 and 45). (Figs. 44–62 are at the end of the chapter, on pages 117–126.)

(b) Maximum bending moments for $p \times 1$, $p \times 3$ and $p \times 5$ grillages. Each of these cases is plotted separately in terms of the distributed reaction parameters (figs. 46–57).

(c) Maximum bending moments for $p \times 7$ and $p \times 9$ grillages. Plotted in terms of the orthotropic plate parameters (figs. 58–61).

In the case of the bending moments, the largest of the values from the electronic computer print-out is usually the one plotted. An exception arises in simply supported beams in grillages having a fairly wide difference in stiffness between the two sets of beams, where two bending moments are required for the less stiff set:

(a) Largest value at an intersection adjacent to the edges of the grillage ($i = 1$).

(b) Largest value at an intersection not adjacent to the edges of the grillage ($i \neq 1$).

In the plotted graphs, both of these are shown where (b) has a larger absolute value than (a), but only (a) is shown once this becomes the larger (at very large or small values of λ, ϵ or μ). In this latter range, (b) may still give a larger total bending moment once the local bending moment has been added, but a safe estimate may be obtained by assuming that (b) is equal to (a). This approximation, which may in a few cases lead to an overestimate of the largest total bending moment, has been introduced in order to simplify the graphs.

11.4. Comparison with earlier data curves by Schade

The results calculated for 9- and 15-beam grillages, and plotted in terms of the orthotropic plate parameters, agree well with Schade's results which were derived by orthotropic plate theory. (see Bibliography, Chapter 3). However, reference to the data curves for smaller numbers of beams shows that the orthotropic plate results may be seriously in error. This is particularly so for the bending moments of the less stiff set of beams in grillages having a wide difference in stiffness between the two sets. Thus, for clamped edges, the orthotropic plate bending moments may be as much as 100 % in error. The errors from Schade's curves for simply supported grillages having a wide difference in stiffness are even greater, since his results apparently give the bending moment at the centre of the grillage, whereas the largest value frequently occurs at an intersection adjacent to the edges.

11.5. Notes on use of data sheets (figs. 44–61)

11.5.1. Choice of boundary condition

The curves have been worked out for opposite pairs of edges either simply supported ($K = 0$) or clamped ($K = 1$). For grillages under uniform pressure which are continuous over bulkheads and joined to similar grillages, the edges should be assumed clamped, due to the symmetry of loading. In other cases, the rotational constraint arises due to the stiffness of the adjoining structure, and edges which are only supported by or attached to a relatively flexible structure should be assumed simply supported. In other cases, judicious extrapolation between the results for $K = 0$ and $K = 1$ might be appropriate.

11.5.2. Number of beams in grillage

The more numerous set of beams are designated the A beams and the less numerous set the B beams ($p \geqslant q$), and specific data curves are available for odd numbers of beams. In each case, the full-line curve gives the deflection or bending moment for grillages containing a large number of beams (p and q large for the deflection, and $p \times 7$ and $p \times 9$ bending-moment curves; p only large in other cases); where the solution differs by more than about 5 % from this value, additional curves are shown as chain dotted lines and labelled with the appropriate beam configuration $p \times q$. Results for numbers of beams not shown may be obtained by interpolation.

11.5.3. Calculation of maximum deflections

The central deflection is obtained from figs. 44 or 45, whichever is appropriate to the end conditions of the A beams. For $p \times 1$ grillages with $K_a = 0$, the result should be multiplied by the factor

$$(0{\cdot}867\bar{q}_a c + 0{\cdot}667\bar{q}_b d)/\bar{p}cd.$$

Occasionally, there may be slightly larger deflections away from the centre due to local sagging between intersections, but this effect is not included.

11.5.4. Calculation of maximum bending moments: general method

The data curves have been calculated with the uniform pressure load considered as an equivalent concentrated load of magnitude $\bar{p}cd$ applied at each beam intersection, and this is often sufficiently accurate for design purposes. An exception commonly arises in grillages where the two sets of beams have markedly different stiffnesses: then, the bending moments of the less stiff set of beams may be significantly affected by local sagging of these beams between the comparatively rigid supports from the stiffer beams at the inter-

[110]

sections, and it is necessary to calculate a term additional to the data sheet result. The local bending term is usually important for the A beams where λ is less than $\frac{1}{5}$ and for the B beams when λ is greater than 5, but, when in doubt, its magnitude should always be checked, even outside these ranges. The additional term is calculated assuming that the pressure on each panel is divided between the four portions of the bounding beams, according to the empirical curve in fig. 42. The bending moment in the individual beam is then calculated for the appropriate loading, assuming that the intersections do not deflect. The local bending-moment diagrams for simply supported edges and for clamped edges are shown in fig. 43, and this moment is added to the data sheet value to give the total bending moment on the beam. This procedure is discussed further below, for different edge conditions. It should be noted that all the data sheet bending moments for $p \times 1$ grillages with $K_a = 0$ should be multiplied by the factor

$$(0 \cdot 867 \bar{q}_a c + 0 \cdot 667 \bar{q}_b d)/\bar{p} c d.$$

11.5.5. Calculation of maximum bending moments in beams with simply supported ends

Case (i) A beams simply supported ($K_a = 0$).
B beams simply supported or clamped ($K_b = 0$ or 1).
Calculation of maximum bending moment in A beams.

The grillage bending moment is obtained from figs. 46, 50, 54, or 58, depending on the number of B beams. In these data sheets, the full line gives the maximum grillage bending moment in the A beams, except where neither set contains more than three beams, in which case the chain dotted line gives the maximum bending moment.

The data sheets also include dashed curves marked C over the range for fairly small ϵ or μ, where local bending of the A beams between intersections may affect the maximum bending moment. The lines marked C give the maximum bending moment in the A beams at the intersection $i = 1$ with the first B beam. Where curve C coincides with the full or chain-dotted lines on the left-hand side of the graphs, the maximum bending moment occurs at $i = 1$, and its value is given by the full or chain-dotted line. Where no curve C is given, the grillage bending moment at $i = 1$ is given by the full line or chain-dotted line throughout.

Towards the right end of the curves (ϵ or μ large), the A beams are the stiffer set, and the maximum bending moment in these beams is given directly by the appropriate full line or chain-dotted curve.

It occurs at the mid-length of a beam but not necessarily in the central beam; as the number of A and B beams and their stiffnesses become more unequal, the position of maximum bending moment moves away from the central beam.

Towards the left end of the curves (ϵ or μ small) the A beams are the less stiff set, and local bending moments have to be added to the data sheet values. The largest total bending moment may arise in one of the following ways:

(a) At an intersection adjacent to the edge of the grillage ($i = 1$). The data sheet moment, which is negative, is reduced by the local bending-moment term: $+0 \cdot 085 \overline{q}_a c^2$ to be added.

(b) At a position midway between the edge of the grillage and the first intersection ($i = \frac{1}{2}$). The data sheet moment is equal to half the value at $i = 1$ given by (a), and it is increased by the local bending-moment term: $-0 \cdot 062 \overline{q}_a c^2$ to be added.

(c) At an intersection other than $i = 1$. The data sheet moment, which is negative, is reduced by the local bending-moment term: $+0 \cdot 067 \overline{q}_a c^2$ to be added.

The largest total arising from (a), (b) or (c) should be used for design calculations, and the procedure is illustrated in fig. 62.

Case (*ii*) B beams simply supported ($K_b = 0$).
  A beams simply supported or clamped ($K_a = 0$ or 1).
  Calculation of maximum bending moment in B beams.

The grillage bending moment is obtained from figs. 47, 51, 55 or 59, depending on the number of B beams. The full lines give the maximum grillage bending moment in the B beams except where there are only a few (generally five or less) beams in each set, in which case the appropriate chain-dotted curve is used.

The data sheets also include dashed curves marked C for fairly large ϵ or μ, where local bending of the B beams between intersections may affect the maximum bending moment. These lines give the maximum bending moment in the B beams at the intersection $i = 1$ with the first A beam. Where curve C coincides with the full or chain-dotted lines on the right-hand side of the graphs, the maximum bending moment occurs at $i = 1$, and its value is given by the full or chain-dotted line.

Towards the left end of the curves (ϵ or μ small), the B beams are the stiffer set, and the maximum bending moment in these beams is given directly by the appropriate full line or chain-dotted curve.

Towards the right end of the curves (ϵ or μ large), the B beams are the less stiff set, and local bending moments have to be added to the data sheet values. The largest total bending moment can arise in one of the following ways:

(a) At an intersection adjacent to the edge of the grillage ($i = 1$). The data sheet moment, which is negative, is reduced by the local bending-moment term: $+0.085\bar{q}_b d^2$ to be added.

(b) At a position midway between the edge of the grillage and the first intersection ($i = \frac{1}{2}$). The data sheet moment is equal to half the value at $i = 1$ given by (a), and it is increased by the local bending-moment term: $-0.062\bar{q}_b d^2$ to be added.

(c) At an intersection other than $i = 1$. The data sheet moment, which is negative, is reduced by the local bending-moment term: $+0.067\bar{q}_b d^2$ to be added.

The largest total arising from (a), (b) or (c) should be used for design calculations, and the procedure is illustrated in fig. 62.

11.5.6. Calculation of maximum bending moments in beams with clamped ends

Case (iii) A beams clamped ($K_a = 1$).

B beams clamped or simply supported ($K_b = 1$ or 0).

Calculation of maximum bending moment in A beams.

Figs. 48, 52, 56 or 60 is used, depending on the number of B beams.

Case (iv) B beams clamped ($K_b = 1$).

A beams clamped or simply supported ($K_a = 1$ or 0).

Calculation of maximum bending moment in B beams.

Figs. 49, 53, 57 or 61 is used depending on the number of A beams.

The procedure in cases (iii) and (iv) is similar. The maximum data sheet moment is obtained from the appropriate full line or chain-dotted curve depending on the beam configuration. This maximum bending moment always occurs at the edge of the grillage but not necessarily in the central beam: as the number of A and B beams and their stiffnesses become more unequal, the position moves away from the central beam.

The effect of local bending is easily allowed for. Since the maximum data sheet bending moment occurs at the fixed end of the beam it is only necessary to add the local bending term $0.067\bar{q}_a c^2$ or $0.067\bar{q}_b d^2$ in cases (iii) or (iv) respectively, to obtain the maximum total bending moment.

11.6. Numerical Examples

Example no. 1: grillage with all edges clamped

5×3 grillage ($p = 5$, $q = 3$) under pressure $\bar{p} = 10\,\text{lb/in.}^2$.

$$K_a = K_b = 1, \quad E = 30 \times 10^6\,\text{lb/in.}^2.$$

Breadth of grillage $a = 12\,\text{ft} = 144\,\text{in.}$
Length of grillage $b = 18\,\text{ft} = 216\,\text{in.}$

$$I_a = 140\,\text{in.}^4, \quad I_b = 60\,\text{in.}^4.$$

i.e. $\quad c = a/(q+1) = 36\,\text{in.}, \quad d = b/(p+1) = 36\,\text{in.},$

$$\lambda = \frac{b^3 I_a}{a^3 I_b} = 7\cdot9, \qquad \epsilon = (p+1)\lambda = 47\cdot4,$$

$$\mu = \frac{(p+1)}{(q+1)}\lambda = 11\cdot8.$$

From fig. 45, for $\mu = 11\cdot8$,

$$w = \frac{0\cdot00283(q+1)\,\bar{p}cda^3}{EI_a} = \frac{0\cdot00283 \times 4 \times 10 \times 36^2 \times 144^3}{30 \times 10^6 \times 140} = 0\cdot104\,\text{in.}$$

From fig. 52, for $\epsilon = 47\cdot4$,

$$M_a = 0\cdot339\bar{p}cda = \frac{0\cdot339 \times 10 \times 36^2 \times 144}{2240} = 282\,\text{tons in.}$$

From fig. 53, for $\epsilon = 47\cdot4$,

$$M_b = 0\cdot0133(p+1)\,\bar{p}cdb = \frac{0\cdot0133 \times 6 \times 10 \times 36^2 \times 216}{2240} = 100\,\text{tons in.}$$

There is an additional bending moment of the B beam due to local bending between intersections. Referring to fig. 42,

$$R = \bar{p}cd = 10 \times 36^2 = 1\cdot296 \times 10^4\,\text{lb},$$
$$R_b = 0\cdot5R = 0\cdot648 \times 10^4\,\text{lb for } c/d = 1,$$
$$\bar{q}_b = 1\cdot5R_b/d = 1\cdot5 \times 0\cdot648 \times 10^4/36 = 270\,\text{lb/in.},$$

and the local bending moment is

$$0\cdot067\bar{q}_b d^2 = 0\cdot067 \times 270 \times 36^2/2240 = 10 \text{ tons in.}$$

i.e. total maximum bending moment of B beams is

$$M_b = 100 + 10 = 110\,\text{tons in.}$$

The maximum bending moments of both A and B beams occur at the edges of the grillage.

[114]

Example no. 2: grillage with one pair of edges clamped and the other pair simply supported

5×3 grillage ($p = 5$, $q = 3$) under pressure $\bar{p} = 10\,\text{lb/in.}^2$. The five beams are clamped at their ends and the three beams are simply supported ($K_a = 1$, $K_b = 0$). $E = 30 \times 10^6\,\text{lb/in.}^2$.

Breadth of grillage $a = 12\,\text{ft} = 144\,\text{in.}$
Length of grillage $b = 24\,\text{ft} = 288\,\text{in.}$

$$I_a = 140\,\text{in.}^4, \quad I_b = 20\,\text{in.}^4.$$

i.e. $\quad c = a/(q+1) = 36\,\text{in.} \quad d = b/(p+1) = 48\,\text{in.},$

$$\lambda = \frac{b^3 I_a}{a^3 I_b} = 56, \qquad \epsilon = (p+1)\lambda = 336,$$

$$\mu = \frac{(p+1)}{(q+1)}\lambda = 84.$$

From fig. 45, for $\mu = 84$,

$$w = \frac{0.00260(q+1)\bar{p}cda^3}{EI_a}$$

$$= \frac{0.00260 \times 4 \times 10 \times 36 \times 48 \times 144^3}{30 \times 10^6 \times 140} = 0.128\,\text{in.}$$

From fig. 52, for $\epsilon = 336$,

$$M_a = 0.330\bar{p}cda = \frac{0.330 \times 10 \times 36 \times 48 \times 144}{2240} = 366\,\text{tons in.}$$

The maximum bending moment of the B beams includes a local bending term. Referring to fig. 42,

$$R = \bar{p}cd = 10 \times 36 \times 48 = 1.728 \times 10^4\,\text{lb},$$
$$R_b = 0.77R = 1.33 \times 10^4\,\text{lb for } d/c = 1.33,$$
$$\bar{q}_b = 1.5R_b/d = 1.5 \times 1.33 \times 10^4/48 = 416\,\text{lb/in.}$$

We may now consider the following three possibilities for the largest total bending moment, using fig. 51 for $\epsilon = 336$:

(a) At an intersection adjacent to the edge of the grillage ($i = 1$).

$$= -0.0014(p+1)\bar{p}cdb$$

Data sheet moment
(from chain-dotted curve) $= -\dfrac{0.0014 \times 6 \times 10 \times 36 \times 48 \times 288}{2240}$

$$= -19\,\text{tons in.}$$

Local bending moment $= +0 \cdot 085 \bar{q}_b d^2$

$$= \frac{0 \cdot 085 \times 416 \times 48^2}{2240} = 36 \text{ tons in.}$$

i.e.   Total bending moment $= -19 + 36 = +17$ tons in.

(b) At a position midway between the edge of the grillage and the first intersection ($i = \frac{1}{2}$).

Data sheet moment = half value at $i = 1$ given under (a)

$$= -9 \text{ tons in.}$$

Local bending moment $= -0 \cdot 062 \bar{q}_b d^2$

$$= -\frac{0 \cdot 062 \times 416 \times 48^2}{2240} = -27 \text{ tons in.}$$

i.e.   Total bending moment $= -9 - 27 = -36$ tons in.

(c) At an intersection other than $i = 1$.

In this instance, the data sheet moment is not specifically given for $i \neq 1$, curve C having merged with the full line curve. The value for $i = 1$ is therefore used as a safe estimate.

i.e.   Data sheet moment $= -19$ tons in.

Local bending moment $= +0 \cdot 067 \bar{q}_b d^2$

$$= \frac{0 \cdot 067 \times 416 \times 48^2}{2240} = 29 \text{ tons in.}$$

i.e.   Total bending moment $= -19 + 29 = +10$ tons in.

Thus, the largest total B beam bending moment arises from (b) and is $M_b = -36$ tons in.

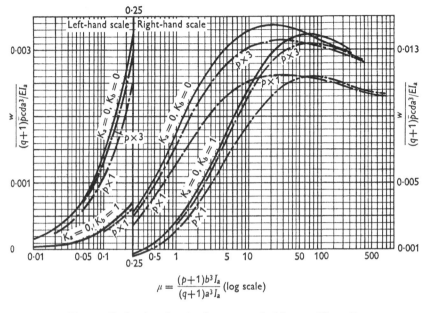

Fig. 44. Deflections for simply supported A beams ($K_a = 0$).

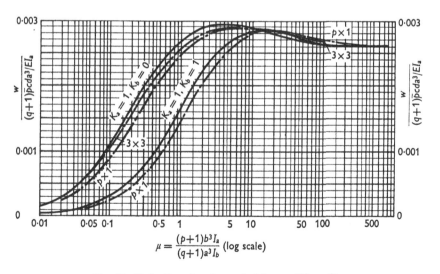

Fig. 45. Deflections for clamped A beams ($K_a = 1$).

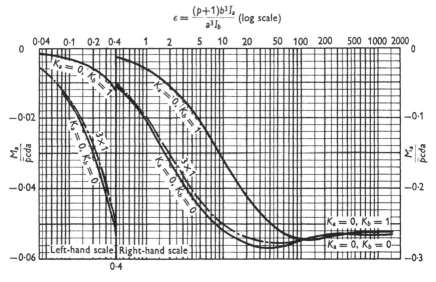

Fig. 46. $p \times 1$ grillages—maximum bending moment in A beams with simply supported ends (M_a for $K_a = 0$).

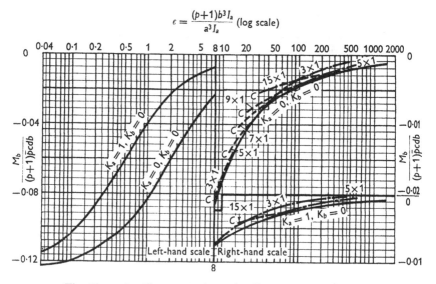

Fig. 47. $p \times 1$ grillages—maximum bending moment in B beams with simply supported ends (M_b for $K_b = 0$).

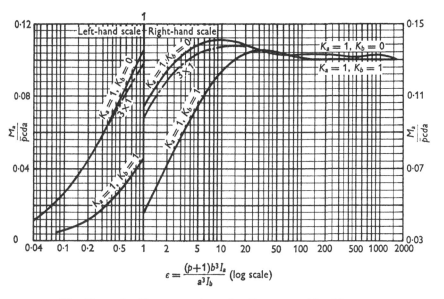

Fig. 48. $p \times 1$ grillages—maximum bending moment in A beams with clamped ends (M_a for $K_a = 1$).

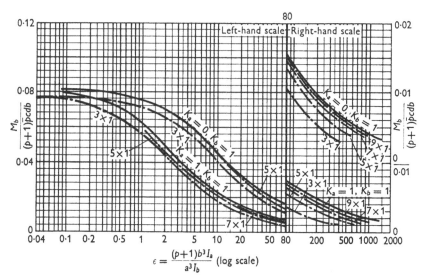

Fig. 49. $p \times 1$ grillages—maximum bending moment in B beams with clamped ends (M_b for $K_b = 1$).

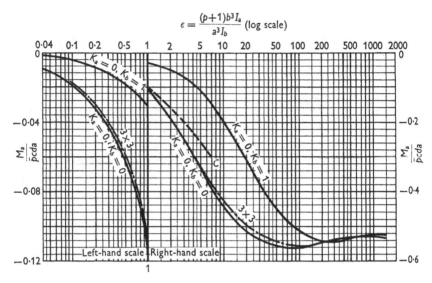

$$\epsilon = \frac{(p+1)b^3 I_a}{a^3 I_b} \text{ (log scale)}$$

Fig. 50. $p \times 3$ grillages—maximum bending moment in A beams with simply supported ends (M_a for $K_a = 0$).

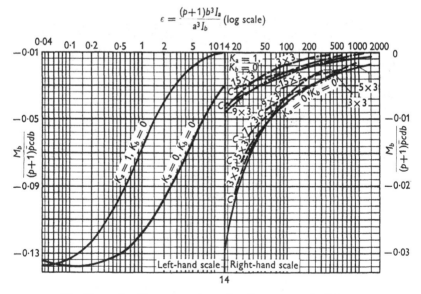

$$\epsilon = \frac{(p+1)b^3 I_a}{a^3 I_b} \text{ (log scale)}$$

Fig. 51. $p \times 3$ grillages—maximum bending moment in B beams with simply supported ends (M_b for $K_b = 0$).

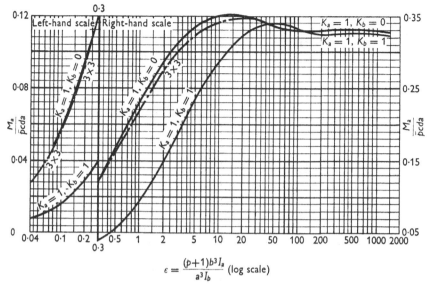

Fig. 52. $p \times 3$ grillages—maximum bending moment in
A beams with clamped ends (M_a for $K_a = 1$).

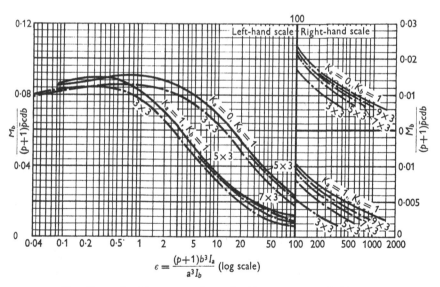

Fig. 53. $p \times 3$ grillages—maximum bending moment in B beams
with clamped ends (M_b for $K_b = 1$).

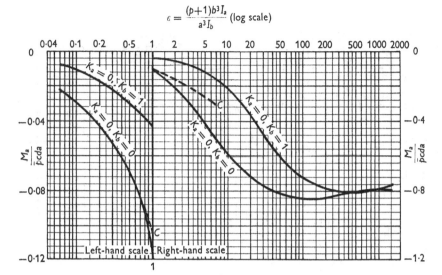

$$\epsilon = \frac{(p+1)b^3 I_a}{a^3 I_b} \text{ (log scale)}$$

Fig. 54. $p \times 5$ grillages—maximum bending moment in A beams with simply supported ends (M_a for $K_a = 0$).

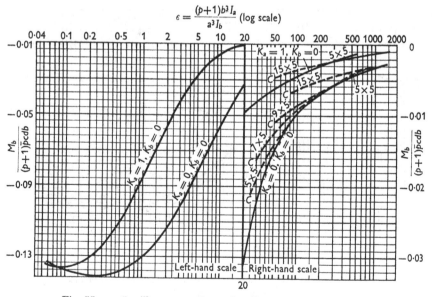

$$\epsilon = \frac{(p+1)b^3 I_a}{a^3 I_b} \text{ (log scale)}$$

Fig. 55. $p \times 5$ grillages—maximum bending moment in B beams with simply supported ends (M_b for $K_b = 0$).

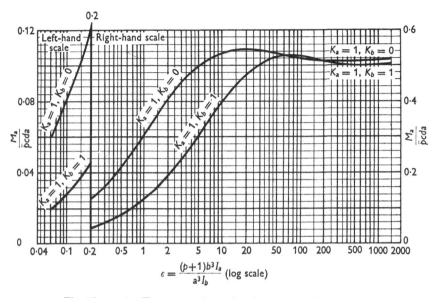

Fig. 56. $p \times 5$ grillages—maximum bending moment in A beams with clamped ends (M_a for $K_a = 1$).

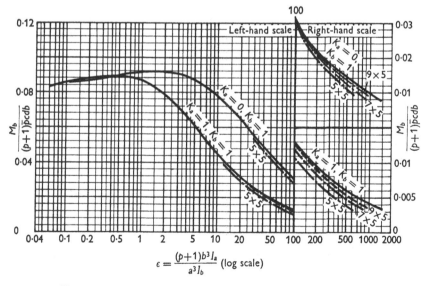

Fig. 57. $p \times 5$ grillages—maximum bending moment in B beams with clamped ends (M_b for $K_b = 1$).

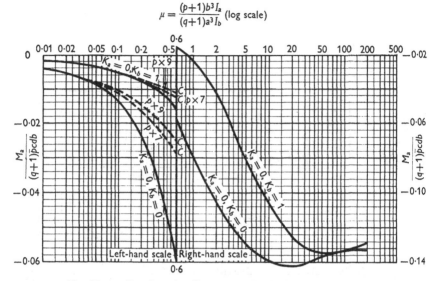

Fig. 58. $p \times 7$ and $p \times 9$ grillages—maximum bending moment in
A beams with simply supported ends (M_a for $K_a = 0$).

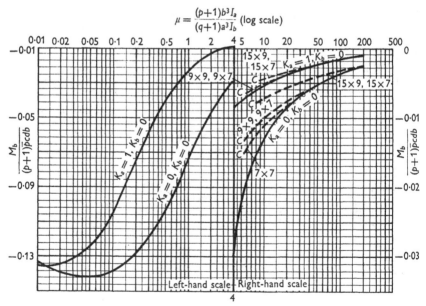

Fig. 59. $p \times 7$ and $p \times 9$ grillages—maximum bending moment in
B beams with simply supported ends (M_b for $K_b = 0$).

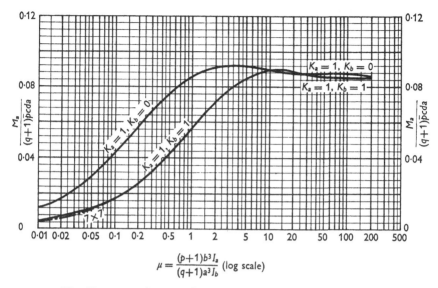

Fig. 60. $p \times 7$ and $p \times 9$ grillages—maximum bending moment in
A beams with clamped ends (M_a for $K_a = 1$).

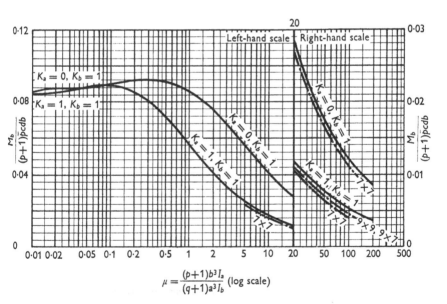

Fig. 61. $p \times 7$ and $p \times 9$ grillages—maximum bending moment in
B beams with clamped ends (M_b for $K_b = 1$).

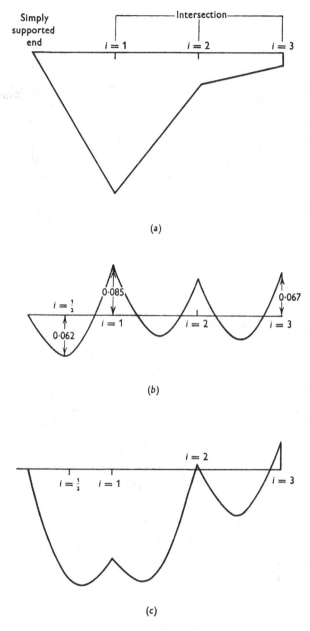

Fig. 62. Procedure for calculating total bending moment for less stiff set of beams. (a) Typical bending moment due to loads at intersection points; (b) bending moment due to local bending (ordinates show approximate values of B.M./$\bar{q}_a c^2$ or B.M./$\bar{q}_b d^2$); (c) total bending moment—maximum value may occur at points such as $i = \frac{1}{2}$, 1, 2, etc.

BIBLIOGRAPHY

Chapter 1

G. Vedeler. *Grillage Beams in Ships and Similar Structures* (Oslo: Grondahl and Son, 1945).

J. F. Baker. *The Steel Skeleton.* Vol. I. *Elastic Behaviour and Design* (Cambridge University Press, 1954).

J. F. Baker, M. R. Home and J. Heyman. *The Steel Skeleton.* Vol. II. *Plastic Behaviour and Design* (Cambridge University Press, 1956).

B. G. Neal. *The Plastic Methods of Structural Analysis* (London: Chapman and Hall Ltd., 1956).

J. Heyman. 'The plastic design of grillages' in *Engineering*, **176**, no. 4587, December 1953, pp. 804–807.

J. B. Martin. 'Large deflection of an encastré rigid-plastic grid' in *J. Mech. Engng Sci.* **4**, no. 4, December 1962, pp. 322–333.

Chapter 2

T. O. Lazarides. 'The design and analysis of openwork prestressed concrete beam grillages' in *Civil Engineering*, **47**, no. 552, June 1952, pp. 471–473.

Chapter 3 (*see also G. Vedeler, above*)

A. W. Hendry and L. G. Jaeger. *The Analysis of Grid Frameworks and Related Structures* (London: Chatto and Windus, 1958).

A. W. Hendry and L. G. Jaeger. 'A general method for the analysis of grid frameworks' in *Proc. Instn Civ. Engrs*, Part III, **4**, no. 3, December 1955, pp. 939–971.

M. Hetényi. *Beams on Elastic Foundation* (Ann Arbor: University of Michigan Press, 1946).

A. J. S. Pippard and J. P. A. de Waele. 'The loading of interconnected bridge girders' in *J. Instn Civ. Engrs*, **10**, no. 1, 1938, pp. 97–114.

H. A. Schade. 'Bending theory of ship bottom structure' in *Trans. Soc. Nav. Archit.*, *N.Y.*, **46**, 1938, pp. 176–205.

H. A. Schade. 'Application of orthotropic plate theory to ship bottom structure' in *Proc. 5th Int. Congr. Appl. Mech.* (New York: J. Wiley and Sons, 1938, pp. 140–144).

H. A. Schade. 'The orthogonally stiffened plate under uniform lateral load' in *J. Appl. Mech.* **7**, no. 4, 1940, pp. A 143–146.

H. A. Schade. 'Design curves for cross-stiffened plating under uniform bending loads' in *Trans. Soc. Nav. Archit.*, *N.Y.*, **49**, 1941, pp. 154–182.

W. H. Hoppmann. 'Bending of orthogonally stiffened plates' in *J. Appl. Mech.* **22**, no. 2, June 1955, pp. 267–271.

P. B. Morice and G. Little. 'Load distribution in prestressed concrete bridge systems' in *Structural Engineer*, **32**, no. 3, March 1954, pp. 83–112.

Chapter 4

J. M. Klitchieff. 'Beams on elastic supports and on cross girders' in *Aeronaut. Quart.* **2**, November 1950, pp. 157–166.

J. M. Klitchieff. 'Beams on cross girders with clamped ends' in *Aeronaut. Quart.* **3**, November 1951, pp. 230–237.

D. F. Holman. 'A finite series solution for grillages under normal loading' in *Aeronaut. Quart.* **8**, February 1957, pp. 49–57.

Chapter 5

D. S. Brooks. 'The design of interconnected bridge girders' in *Civ. Engng Publ. Works Rev.* **53**, nos. 623 and 624, 1958, pp. 535–538 and 682–684.

E. Lightfoot and F. Sawko. 'Structural frame analysis by electronic computer' in *Engineering*, **187**, no. 4843, January 1959, pp. 18–20.

Chapter 6

S. Kendrick. 'The effect of deflexion due to shear on the stresses and deflexions of a plane grillage of beams' in *Proc. Instn Civ. Engrs*, Part III, **5**, August 1956, pp. 527–531.

Chapter 7

S. Kendrick. 'The analysis of flat plated grillages' in *European Shipbuilding*, **5**, no. 1, 1956, pp. 4–10.

J. Clarkson. 'Tests of flat plated grillages under concentrated loads' in *Trans. Instn Nav. Archit.* **101**, no. 2, 1959, pp. 129–142.

J. Clarkson. 'Elastic analysis of a beam-plating structure under a single concentrated load' in *Proc. IX Int. Congr. Appl. Mech.*, *Bruxelles*, 1956, pp. 176–186.

J. Clarkson. 'The behaviour of deck stiffening under concentrated loads' in *Trans. Instn Nav. Archit.* **104**, no. 1, 1962, pp. 57–65.

J. Clarkson. 'Tests of flat plated grillages under uniform pressure' in *Trans. R. Instn Nav. Archit.* **105**, no. 4, 1963, pp. 467–484.

Chapter 8

D. Faulkner. 'Welded connections used in warship structures', in *Trans. R. Instn Nav. Archit.* **106**, no. 1, 1964, pp. 39–70.

Chapters 9 and 10

J. Clarkson. 'The design for minimum weight of simply supported flat grillages to withstand a single concentrated load' in *Trans. N.E. Coast Instn Engineers and Shipbuilders*, **73**, part 3, 1957, pp. 145–178 and D 49–51.

J. H. Lamble and Li Shing. 'A survey of published work on the deflection of and stress in flat plates subject to hydrostatic loading' in *Trans. R. Instn Nav. Archit.* **89**, 1947, pp. 128–147.

J. Clarkson. 'A new approach to the design of plates to withstand lateral pressure' in *Trans. R. Instn Nav. Archit.* **98**, no. 4, 1956, pp. 443–463.

J. Clarkson. 'The strength of approximately flat long rectangular plates under lateral pressure' in *Trans. N.E. Coast Instn Engineers and Shipbuilders*, **74**, part 1, 1957, pp. 21–40 and D 1–8.

J. Clarkson. 'Uniform pressure tests on plates with edges free to slide inwards' in *Trans. R. Instn Nav. Archit.* **104**, no. 1, 1962, pp. 67–80.

J. E. Greenspon. 'An approximation to the plastic deformation of a rectangular plate under static load with design applications' in *International Shipbuilding Progress*, **3**, no. 22, June 1956, pp. 329–340.

A. G. Young. 'Ship plating loaded beyond the elastic limit' in *Trans. R. Instn Nav. Archit.* **101**, no. 2, 1959, pp. 143–165.

Chapter 11

J. Clarkson, L. B. Wilson and J. L. McKeeman. 'Data sheets for the elastic design of flat grillages under uniform pressure' in *European Shipbuilding*, **8**, no. 8, 1959, pp. 174–198.

INDEX